Woodlands Management Course

A Guide to Improving Our Forests

900 Circle 75 Parkway, Suite 205
Atlanta, Georgia 30339

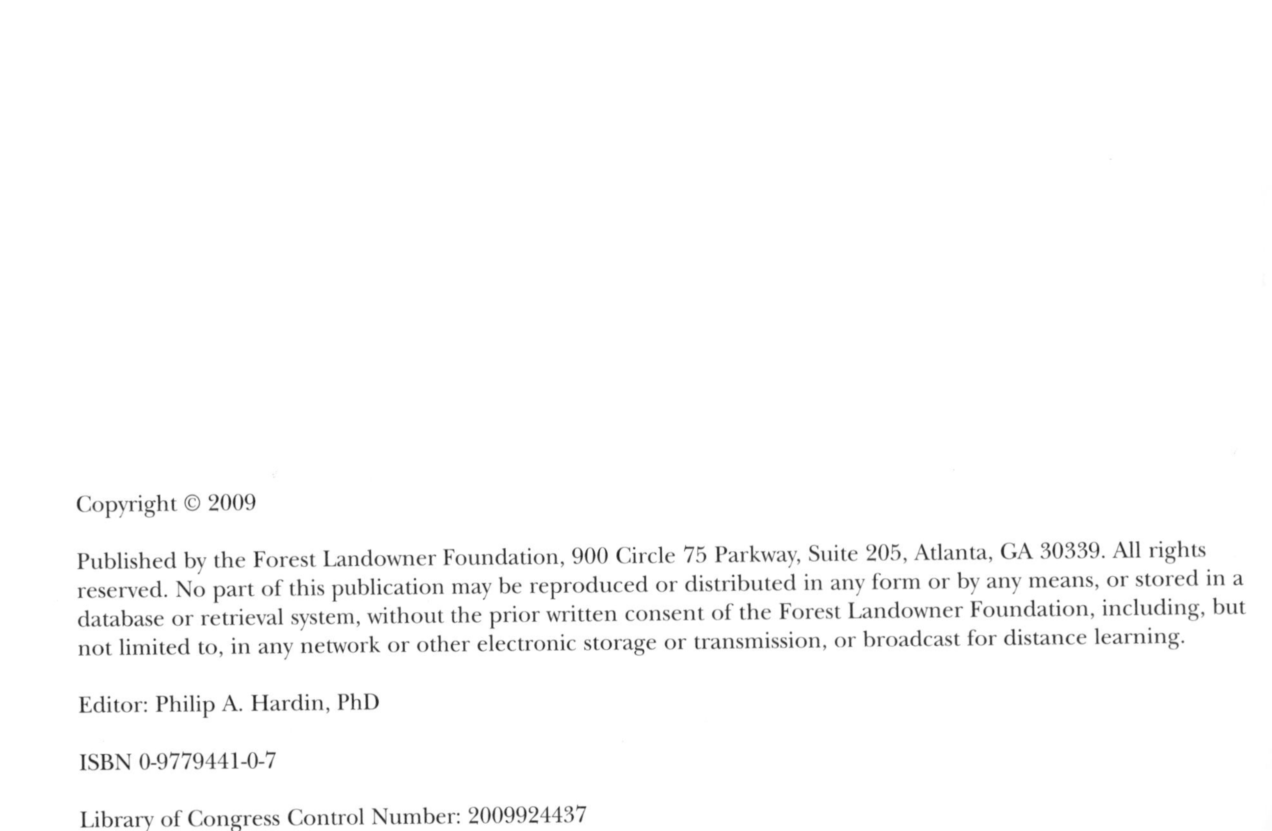

Published by the Forest Landowner Foundation, 900 Circle 75 Parkway, Suite 205, Atlanta, GA 30339.

Editor: Philip A. Hardin, PhD

ISBN 0-9779441-0-7

Library of Congress Control Number: 2009924437

To our children, our grandchildren and the future generations.
It is for them that we plant the trees and nurture the forest.

Woodlands Management Course:

Table of Contents

Chapter Authors and Editors Credits

The first version of this book was published in 1985 with the title "Woodland Management, a Correspondence Course for Forest Land Owners," and was developed by Extension Forest Resources Department, School of Forest Resources, N.C. Agricultural Extension Service, N.C. State University, Raleigh, N.C. The book was edited by Gary B. Blank, and its publication was sponsored by the Forest Farmers Association, Inc. (The Forest Landowners Association prior to the name change) and the Cooperative Extension Services, Southern Region. You can obtain a copy of this original version from the Southern Regional Extension Forestry website at:
http://sref.info/publications/online_pubs/woodland_correspondence.pdf/file_view

In the mid 1990s, the Forest Landowners Foundation raised funds to update the publication, and using a portion of these funds, paid for an update and enlargement of the publication under the direction of Dr. Dick Brinker at Auburn University and Mr. Rick Hamilton at N.C. State University.

After the text was updated, the book, unfortunately, was "put on the back shelf" due to changes in personnel at the Forest Farmers Association. In the mid 2000s, the partially completed work was re-discovered, and the Forest Landowner Foundation began work to complete the publication. Since a number of years had passed since the latest edits, the first step was to have the text reviewed and re-edited. A new group of editors were contacted with the assistance of Bill Hubbard, the

Southern Regional Extension Forester. These new editors updated and enlarged the text so that it was relevant to current forest ownership concerns. They are credited in the section below.

At the same time the last round of edits were being made, the decision was made to make the publication a reference book, and move away from the correspondence course form. The reasons for this decision included the difficulty in finding people to grade correspondence course submissions, the difficulty of administration, and the high cost to the students. Our objective was to have a widely available resource that could benefit as many landowners as possible.

In the following credits, there are a number of descriptions of the amount of work done by the latest set of chapter authors and editors, "edited", "authored", "revised", etc. These reflect the amount of work contributed, and that was determined by the change in knowledge over the time the text had remained dormant. All of the authors were generous with their time, their knowledge, and their talents, and we are grateful to all of them for their effort and cooperation. Without them, this book could not have been published.

Chapter 1—Why Do I Own Forest Land?

Extensively revised and updated by:
Mr. Kris Irwin, Public Service Associate
Daniel B. Warnell School of Forest Resources
The University of Georgia
Athens, GA

Chapter 2—Why Have a Management Plan?

Edited by:
Mr. William G. Hubbard, Southern Regional Extension Forester
Cooperative Extension Service
The University of Georgia
Athens, GA

Chapter 3—Managing for Wildlife.

Edited by:
Dr. Greg K. Yarrow, Extension Specialist
Clemson University
Department of Forest Resources
Clemson, SC

Chapter 4—Nature-based Tourism Opportunities for your Woodland

Revised and updated by:
Dr. Taylor Stein, Assistant Professor - Forest Recreation
The University of Florida
Gainesville, FL

Chapter 5—Forest Soils and Site Index

Edited by:
Dr. James E. Johnson, Associate Dean, Outreach
Virginia Tech
College of Natural Resources
Blacksburg, VA

Chapter 6—Southern Pine Reforestation

Revised and updated by:
Dr. Kenneth L. McNabb, Extension Specialist
Auburn University
School of Forestry
Auburn University, AL

Chapter 7—Regeneration of Southern Hardwoods

Edited and updated by:
Dr. Wayne Clatterbuck, Associate Professor
University of Tennessee
Dept. of Forestry, Wildlife & Fisheries
Knoxville, TN

Chapter 8—Finance, Taxes and Investment Issues

Extensively revised and updated by:

Mr. William G. Hubbard, Southern Regional Extension Forester
Cooperative Extension Service
The University of Georgia
Athens, GA

and

Ms. Deborah A. Gladdis
Associate Extension Professor
Department of Forestry
Mississippi State University
Starkville, MS

Chapter 9—Managing Healthy Forests: Prevention and Control of Major Forest Insects and Diseases in the Southeast

Authored by:

Dr. Dennis Hazel, Assistant Professor and Extension Specialist
Mr. Rick A. Hamilton, Extension Specialist and Department Extension Leader, retired
Dr. Mark A. Megalos, Assistant Professor and Extension Specialist
Department of Forestry and Environmental Resources
North Carolina State University
Raleigh, NC

Chapter 10—Intermediate Pine Stand Management

Edited and reorganized by:

Dr. David J. Moorhead, Extension Specialist
University of Georgia
Tifton, GA

Chapter 11—Intermediate Hardwood Stand Management

Extensively expanded and revised by:

Dr. Jeffrey W. Stringer, Extension Specialist
University of Kentucky
Department of Forestry
Lexington, KY

Chapter 12—Marketing Your Timber
Edited by:
Dr. Mike Dunn, Coordinator, Extension Natural Resources Program
Louisiana State University
LSU Ag Center
Baton Rouge, LA

Chapter 13—Federal Environmental Regulations and the Private Forest Landowner
Edited and with extensive updates by:
Dr. Ben D. Jackson, Extension Specialist
University of Georgia
Cooperative Extension Service
School of Forest Resources
Athens, GA

Chapter 14—Private Property Rights and Potential Liabilities
Edited and updated by:
Dr. Harry L. Haney, Jr., Garland Gray Professor Emeritus of Forestry
Department of Forestry
Virginia Tech
Blacksburg, VA

Chapter 15—Forest Sustainability and Certification
Authored by:
Mr. L. Keville Larson
Larson & McGowin
Mobile, AL

Chapter 16—Being Informed, Taking Action
Authored by:
Mr. Scott P. Jones, Executive Vice-President
The Forest Landowners Association, Inc.
Atlanta, GA

Appendix H—How to Choose a Consulting Forester

Authored by:
Mr. L. Keville Larson
Larson & McGowin
Mobile, AL

This appendix is taken, in its entirety, from Forest Landowner *magazine, January/February 2003, and is used with permission.*

The Unit Headers were all revised by:
Mr. William G. Hubbard, Southern Regional Extension Forester
Cooperative Extension Service
The University of Georgia
Athens, GA

Sponsor Recogniton

We gratefully recognize the sponsors who, through their generous donations, made this book possible.

We are especially grateful to the US Forestry Service for their particularly generous contribution.

The individuals who made donations, Drs Shelton and Jean Short are particularly thanked. We appreciate their assistance in making this book possible.

U. S. Forest Service
The American Forest & Paper Association
Alabama River Woodlands
Weyerhaeuser Company Foundation
Westvaco Corporation
Champion International
Southeastern Lumber Manufacturers Association, Inc.
Georgia-Pacific Corporation
Drs Sheldon & Jean Short

Introduction

This book is designed for landowners and others who are looking for current woodlands management information. It is not designed for professional foresters, nor is it designed to make the layperson a professional forester. Many of today's landowners have a full-time "day job" or other interests that prevent them from spending full time managing their forests, and they do not have the time to find or read in-depth technical papers; however, they want and need more knowledge about the forest, and they prefer it to be presented in an easy to understand format. To that end, this book has been produced to meet the following objectives:

- to let landowners know the options for managing their forests;
- to make landowners aware of the benefits and costs, both monetary and non-monetary, of forest management, and to introduce them to ways to analyze and predict these benefits and costs;
- to assist landowners in defining their objectives, in organizing activities to meet those objectives, and in finding professional foresters and others to manage the activities;
- to generate further interest in the management of our nation's forests.

This book covers two subjects, forest management and the laws and politics of forestry. The forestry management sections include topics not only about managing forests for their timber, but also for wildlife, recreation, water quality, simple enjoyment, etc. The first chapters cover goals, objectives, and planning. Wildlife and timber management are covered in the middle chapters, and timber harvests and sales are addressed toward the end. The last four chapters are devoted to the regulatory, legal, and political aspects of forest land. A glossary defines words and concepts that might not be familiar. Information on specialty tree crops such as Christmas trees, pecans, peaches, etc. is not included as their management is largely influenced by their specific location and, in addition, there is generally a wealth of resources available from local organizations.

The chapters on laws and politics are included because, as much as we might wish, we cannot escape the politics of the forests. Public attention and media focus require that we be involved. To do a good job managing of our forests, we must be knowledgeable about the current political climate, and we must be active in influencing changes in the laws and regulations that govern our forests. Many of our forestry regulations do improve the long term viability of the forests and "level the playing field" for those of us who manage for the long term. Unfortunately, some regulations contain requirements instituted because someone thinks they are a good idea, but, all too often, are detrimental to both the forests' health and the objectives stated as the reason for the requirements. All forest owners need to actively work to ensure that regulations and laws will improve forest health, and, moreover, will encourage owners to improve their forests. Being an active member of the Forest Landowners Association will enable you to have an effective voice at the national level. You should also participate in local and state forestry organizations to have influence at those levels. Forest owners need an advocate at every level of government where they are regulated or taxed.

Information on managing for the newest markets such as carbon sequestration, carbon offsets, or ecosystem benefits (eg, clean water and air) is not included as these concepts are new and very much in flux. Many of the reference sources noted at the end of the chapters will have current information on these subjects. In addition, several forestry publications are following these topics, and you can stay abreast of the latest developments by reading the articles as they appear. Participating in a forest owner organization, especially the Forest Landowner Association, will assure that you are kept up to date. With all of the enthusiasm about these new markets as possible revenue sources, one is cautioned not to become too involved before all the benefits and, more importantly, the costs are reasonably well defined.

For the most complete understanding of the material in this book, reading it from front to back is recommended. However, each chapter is designed to be self contained, so you can read only those chapters of interest at any time. But, if you read only a few chapters, you will miss the interrelation of that material with the remainder of the book. In any event, you are encouraged to be familiar with the sections on planning to help you determine your objectives and to build an overall management plan before you start your forest management activities.

The material in this book is not meant to exhaustively cover each subject addressed, but is designed to be a thorough introduction and a good familiarization with the concepts. Even though the chapter editors and authors are some of the most highly qualified in their areas, for your in-depth understanding, and particularly for detailed considerations in specific circumstances and local conditions, more research and investigation will be required. In addition, as the time between this book's publication and your reading increases, the information has a greater chance of becoming outdated. This is especially true of the chapters that deal with regulations, laws and the government incentive programs.

To obtain up-to-date information or more detailed reference material, each chapter contains a list of resources and the appendices contain references to selected organizations that can be of assistance. Also, libraries and university archives are an especially good source for additional information. Although the internet is an accessible, easy to use reference resource, you have to use caution as the material you find is often superficial, is sometimes incorrect, and is often outdated. (For some reason, determining the publication date of much of the information on the internet seems to be especially difficult.) A forestry professional or a library at one of the forestry schools listed in the appendices can provide a list of current references.

After reading this book, you will have a better understanding of the needs of your forest, and you can formulate a plan for your land. It is recommended that you have a qualified forestry professional assist you as they will have more in-depth knowledge and can provide essential local information that this book does not include. (The cost of their services, will, as many studies have shown, be repaid many times over.) After you have written a draft of your plan, look it over carefully, and

- Review your goals; are they really what you want?
- Review the key steps and examine each one carefully. Will the steps achieve your goals? Are there other management steps you could take to better manage your forests or to achieve results faster or at a lower cost?

After you and your forestry consultant are fully satisfied with your plan, actually carry out the forestry work. And, make sure the work is completed to specifications.

Perhaps, as you formulate your plan, you will have more questions than when you started, but, more than likely, the questions will be focused and will concentrate your attention on the more critical aspects of managing your forest land. In spite of having remaining questions, move ahead to construct your forestry plan and make a note of the questions so you can answer them later. Many studies have shown that, even if you do not have the best possible plan, managing forest lands using validated practices yields much, much better results than no management or work.

We know our world is dynamic and new information is always available, and we know that forests do not always develop as planned. Every few years, review your plans and the condition of your forests, and allow for mid-course corrections. Incorporate any needed changes and any newer forestry practices (making sure these are both verified and pertinent) into your forestry plans and activities. The important steps are to perform a review every few years, to make the changes, to incorporate the changes into your forestry work, and then to carry out your planned work.

Whether you own a few acres or several thousand acres, you will find this book helpful. As one of our landowner reviewers commented "this book has the information about managing my forests that

I have been trying to compile over the last fifteen years." No matter what the condition of your forests, they will be improved by applying the knowledge in this book.

As an important part of owning forest land, remember, while you are working hard to improve your land and your trees, to take the time to enjoy the varied aspects of your forests.

Philip A. Hardin, Ph.D.
Editor

Past President, Forest Landowner Foundation
Bellevue, Washington

Unit I:

Forest Management Planning

Forests occupy an important place in our life, but relatively few people understand very much about wooded land, its value to our well-being or *our* impact on its well-being. In fact, many people tend to take forests for granted and ignore their condition until the time comes when they would like something from the forest—wood and paper products, wildlife, recreation, clean water or solitude.

A key to better understanding and management of the forest is the comprehension of the natural forces at work. Forests are dynamic systems, comprised of plant and animal communities, which interact with each other and their environment in complex ways over time. Soils and nutrients change with forest development. Species of trees and other plants will age and give way to different species. Animals dependent on certain plant species will leave when those species are replaced by different ones. New animal species will move into the area and, in turn, be replaced by still other animal species as the plant community continues to change.

The natural order of change is termed forest succession, meaning that stages of development occur in a progression towards a climax state. This climax state is a relatively stable condition in which change still occurs but at a slower rate. The dynamics of forest growth are controlled by natural laws that, in the absence of human intervention, will surely alter conditions in the forest. Even the most remote forest, those least affected by human activity, is changing continually. These changes are rarely recognized by the casual observer.

The history of forest use and abuse reflects the effects that human intervention can cause. Too often, human interruptions in the natural processes of forests have been destructive because people did not understand or did not care about the natural laws governing forests. But it is illogical to assume that no interruptions in natural forest processes should occur and that the best forests are always those that are left untouched. Natural catastrophes, diseases, insects, and wildfires all can harm forests and may be apt to occur in those areas where people have had the least impact. As natural parts of the environment, these harmful influences are inevitable unless something, usually human intervention, happens to limit the forest's susceptibility to harm.

Forest management is an active process. Humans use the forest in productive ways to meet their needs by altering natural succession through harvesting, regenerating and tending trees. Owners of woodlands who become involved with their lands and develop plans for managing their forest resources will enjoy the most benefit from their forest. They should base their plans on thoughtful consideration of the land itself and important facts concerning their families.

Unit One examines those factors that underlie and affect your choice of management objectives. It also looks at the characteristics of different management plans, so you will know what to expect and be prepared to deal effectively with those who are trained to assist you in managing your forest resource.

Chapter 1:
Why Do I Own Forestland?

Owning forestland can be a great adventure! It can provide you many returns—both financial and personal. If you own forestland now, or plan on buying a tract of land in the future, ask yourself one question, "What am I going to do with it?" Do you want to hold it for the investment value? Do you also want to use the land for recreation (i.e. hunting, fishing, hiking, camping, etc.)? Or maybe you want to grow and harvest timber for additional income. These are all valid reasons for owning forestland, and there are others that we will discuss.

This chapter will cover the need for selecting management objectives, and address a few of the personal issues you will want to consider in your plan.

Objectives of Ownership

An objective is defined as "a goal or end of action; or something toward which effort is directed." The focus of this chapter is to guide you in selecting the objectives best suited to meet *your* needs. Knowing what you want from your land is a critical first step in the planning process. The plan is *your* road map to achieving *your* desired results.

At some point during the first few conversations you have with your professional forester, you are going to be asked what your objectives are for your forestland; what do you want to do with, or gain from, your forestland? By asking this question the forester is seeking direction. Your response will serve as a guide to devise a management plan best suited for you and your land. The more information you can articulate, the more precisely a plan can be created to satisfy *your* vision.

The objectives you choose need to be realistic. They should be based on the current condition of the forest and natural characteristics of the land, and the circumstances in your life, too. You should manage your land in such a way that you receive the benefits, directly or indirectly. Benefits include things like income, clean water, resource conservation, recreation, and personal enjoyment.

As you focus on your objectives, keep in mind the realities of fluctuating market demands and prices for timber and non-forest products. The prices for timber vary from region to region, state to state, and county to county. Consulting foresters keep up with the current market prices for timber products like pulp, chip-n-saw, sawtimber and poles. Non-forest products include wildlife, recreation, clean water, edible plants, and even solitude. You might also want to consider talking with adjacent property owners. Find out what their plans are, because whatever they do may have either a positive or negative effect on your current and future activities.

It is highly advisable to work with a professional forester. You can hire a consulting forester or contact your state forestry agency to find out what services they will provide you. A forester can provide information and explain the details regarding current market trends, site characteristics, and evaluate the current condition and value of the trees and other resources existing on your land. However, for a forester to accurately perform the tasks you desire, certain information that can come only from analysis of your own situation must be supplied. Therefore, you must decide the objectives you have for owning the land.

As you read and work your way through this section of the chapter, consider these basic questions. You may think of some additional questions of your own.

- Do you require income from your forestland before you can afford to implement any management practices?
- Are you willing to invest money that may not provide any financial return?
- Are you willing to perform some of the management activities yourself?
- What would you like to have happen to your forestland after your death?

Management Options

Trees can grow and die without any intervention from humans. They grow wood, provide wildlife habitat, protect and add organic matter to the soil, produce oxygen as a byproduct of their food factory (leaves and photosynthesis), and diversify the vertical composition of the landscape. This is true of most any forest, whether it is managed or not. However, as the owner of forestland, you may want to consider managing your forest to produce timber and/or provide a place for your family to enjoy themselves.

Whatever your reasons for owning forestland—know one thing—the goal of a managed forest is to keep it healthy and productive. Management activities to maintain forest health include removal of dead and dying trees and trees that are struggling to grow (thinning); harvesting of mature trees to generate income and diversify the composition of the forest species; construction of roads and trails to provide access throughout your property; creation of ponds or other impoundments to provide recreation and serve as a source of water in the event of a fire. A healthy forest is better able to resist attack from insects and diseases. Another part of management is providing protection from wildfire (lightning strikes and arson). This includes the installation of firebreaks to keep fire out, and contain any fires that do start in a small area. Yes, your management scheme might require the use of controlled burning to reduce the buildup of fuel (leaves, branches, twigs, grasses, etc.) that can, if accidentally ignited, destroy your forest. It does not matter if you are growing trees for profit or just for fun, proper management techniques will benefit your forest.

To obtain the greatest benefit (tree growth, beauty, wildlife habitat, income, etc.), it is vital that you decide what you want from your forest. Whatever your choice, identifying the objectives is the first step in direction of developing a management plan. You will learn more about management plans in Chapter Two.

You can apply management activities to achieve a single objective or to meet the needs of multiple objectives on a single tract of land. Objectives can change and be modified; they are not carved in stone. As you increase the number of objectives, the more carefully your plan has to account for details. Sometimes the same practices produce different results—it depends at what stage in the forest's life that the practice is implemented. The following are brief descriptions of the most common objectives for forestland.

Timber Management

Do you want to grow your trees to produce timber? Timber production is a common objective for many forest landowners. Timber is a sustainable and renewable resource that can provide income to you and your family. Timber is a long-term objective that requires a detailed management plan, and is certainly compatible with other objectives like wildlife management, aesthetics, and recreation.

Timber management includes activities such as regeneration of the new forest after mature trees are harvested, intermediate treatments (thinning, competition control, pruning, etc.) to maintain tree health and vigor and increase wood quality, road construction (permanent and temporary), inventory of the standing timber (timber cruise), and harvesting.

There are direct and indirect costs associated with many of these activities, but opportunities exist to recover your costs, too. Thinning is a management activity that can produce income when the trees harvested have commercial value. This is referred to as a commercial thinning. There are other sources of income from your timber that do not require any trees to be harvested; for example, collection of the pine needles, commonly known as "pine straw," is a sustainable source of income. The amount received depends on the tree species. Needles from the longleaf pine bring top dollar, but straw from slash pine and loblolly pine is also raked, baled and sold.

Firewood Management

Wood, as a fuel, is available from any tract of forestland. However, not all trees are created equally. Wood from some tree species is preferred over others because of its BTU value (British Thermal Unit), which refers to the quantity of heat produced from a given amount of wood. Other factors that impact the quality of wood are moisture content, ease of splitting, and ash content.

Firewood production can be integrated with many other forest management objectives especially where partial cutting to improve wildlife habitat and stand quality is desired. It is advisable *not* to allow visitors on your land to harvest the trees, buck the logs, split, or even stack the wood. You are asking for a liability lawsuit if someone gets hurt. In most cases, it is a job you will do, or you will contract the work to be done.

Selling firewood can be profitable depending on your location and product demand. Prices are typically based on a volume of wood called a "cord." A cord of firewood contains 128 cubic feet (a stack that is four feet wide by four feet tall by eight feet long). The price depends on the species and quality of the wood. Is it dry or still green? Is it split or not? Are you going to deliver or will the buyer pick it up? If you have a market for firewood in your area, inquire about the going price for a full cord and half cord volumes. It might prove to be a profitable endeavor.

Wildlife Production

Forests provide food, shelter, and water for wildlife species. Some are valued as game species (regulated harvesting) while others are classified as non-game species. No matter what species are desired, there are three fundamentals associated with wildlife management: habitat, wildlife population, and human behavior. Exercising some control over each factor will determine the number and frequency of species on your property. Keep in mind that there are laws regulating the harvesting of game species and providing for the protection of the non-game species. Your state wildlife agency can give you information about these laws.

Hunting leases may be a source of extra income for you. A lease of this nature is a legal binding agreement between the lessee (usually a group of hunters) and the lessor (landowner) for the exclusive right to hunt a piece of land for a given amount of time. The price paid for the lease is most often based on a per acre rate. The amount varies, but is typically based on the target species (deer, waterfowl, turkey, quail), availability of facilities, and quality of hunting. As a "rule of thumb", the fees charged for leasing land for hunting should be equal, or greater than the amount of your property taxes. For instance, if your forestland is conveniently located to a major metropolitan area and you have quality habitat that support abundant numbers of game species, you can charge a price per acre that exceeds your annual property taxes. Basically, it is simple economics of supply and demand. Because land available for leasing is limited and the number of people looking for land to lease for hunting is increasing, then the price you charge is set by you, the forest landowner.

If you are interested in leasing your land for hunting there are a few things to think about—the wildlife assets, the natural characteristics of your land, and the people who will lease your land. Does your land have the appropriate habitat to support the target species? Do the people who lease your land agree with your ideas and goals for your land? Whatever you do, buy insurance, have a written contract, and talk with a wildlife biologist. The insurance is legal protection for you, and the biologist can help you manage your land to attract wildlife.

Recreation

Forestland can be a place for family leisure activities like camping, horseback riding, and hiking. Such activities often require manipulation of the landscape such as construction of trails, clearing for camp sites, and mapping of trails. Once completed, these add value to your experience, and to the resale value of your land. However, care must be taken to protect the soil resources from high impact activities such as horseback riding and all terrain vehicles (ATVs) that can cause soil erosion and compaction if trails are not properly designed, installed, and maintained.

If your land harbors unique habitats you might open it to public access. If this is the case, you should seek legal counsel to make sure you have the appropriate protection against liability claims. Another idea is to develop multiple recreation opportunities (bike trails, wildlife photography, ecotourism) and charge a user fee; this could be profitable, particularly in areas lacking public access lands, but you will need legal advice.

Recreation and hunting usually do not mix well. Therefore, if you wish to do both, each will have to be done on a seasonal basis. But recreation can be combined with timber and non-timber production, aesthetics, or as a stand-alone objective.

Aesthetic Enjoyment

Many landowners are simply interested in the visual value of the land. In such a case, you must be able and willing to absorb the costs of ownership (taxes, road and trail maintenance, etc.) and forego possible profits from timber harvesting, hunting leases, or user fees. Managing your forestland for aesthetic enjoyment, however, does require the forest be protected and maintained.

Individual or groups of trees might need to be harvested to enhance the natural beauty of your forest. These trees may or may not be of economic value, but that will not matter if your objective is to increase the aesthetic value and nothing else. Trees infested with disease and insects must be identified and removed quickly before they create additional problems. Planting native flowering trees and shrubs is easy and inexpensive, and adds diversity to your forest, too. On the other hand, you may find you have to remove invasive species such as privet and kudzu. Other ideas include establishment of wildlife food plots and viewing blinds. If you are holding the land for eventual real estate development, attention to the aesthetic quality will increase its value. As an objective, aesthetic enjoyment fits with most any other objective, even occasional and partial timber harvests.

Watershed Protection

A watershed is defined as a specific land area that drains water into a river system or other body of water. Forested watersheds protect and improve water quality in many ways. A forested landscape acts like a natural sponge; it absorbs water and releases it slowly. As the water passes through the system it is also filtered. The forest cover (leaves, twigs, branches, and trunk) intercepts precipitation (rain and snow), and effectively slows the water movement, thus allowing water more time to infiltrate into the soil. Forest soils are very absorbent because they are composed of decaying leaves, twigs, branches, tree trunks, and other material that hold the water. Once in the soil, some of the water is absorbed by plant roots and transported through the plants to the leaves where it is released back into the atmosphere (transpiration). The rest of the water continues down into the subsoil where it comes in contact with subsurface storage (ground water) or it may reach deeper areas known as aquifers. The bottom line is that a forested watershed yields the greatest amount of clean water returned to the water cycle. Municipalities are recognizing more and more the value of nearby forests. One of the best sources of drinking water in the United States is the Catskill Mountains, which supply excellent drinking water for New York City.

Timber management activities have a short-lived, minimal impact on water quality when properly executed. High water quality can be maintained with other objectives, including timber production. Protecting the integrity of a watershed is of primary importance and is accomplished by following guidelines commonly referred to as Best Management Practices or BMPs. They minimize the impact on water quality from soil disturbances that occur during forestry activities such as harvesting, road/trail construction, stream crossings, and site preparation before planting. Your state forestry agency can give you details about the BMPs applicable in your state.

Future Development

Your forestland may eventually be a prime location for residential, recreational or industrial development. Urban development is spreading rapidly. What is now considered a rural area may become a suburban area within a few years. As such, your land will increase in value due to economic factors, whether you manage it or not. Therefore, depending on the current development trends near your forestland, you may profit from judicious management by enhancing it to meet the needs of future development (especially for residential development). For example, wetland restoration activities (returning previously drained land back to a functional wetland) can increase the value of your property. Keep in mind, you own the land and have the right to manage and use the land in any way you wish. Keep your objectives in mind, but do not forget to pay attention to the development activities in the areas surrounding your property.

Concerns of Ownership

Land ownership has played a major part in the shaping of our nation. Current ownership patterns reflect changes in our society as we have moved from rural to urban and suburban life styles. Our highly mobile society has dispersed family members across the country. Your heirs may live far from the land they stand to inherit one day, while others may still reside on the family land or live nearby. Longer life spans combined with increased income potential affect land ownership and management opportunities as well. Many factors affect the motivation and ability to own forestland.

Ownership Status

Landowners can be classified by the size of their holdings, financial status, occupation, age, and ownership status. Classifying landowners helps foresters identify the types of management programs and practices needed for each individual landowner. Think about your situation. Do you live a few miles from your property, in the adjoining county, or in another state? If you live on or near your forestland then you are considered to be a resident landowner. On the other hand, if you live some distance from your property, perhaps in another city or state, you are considered an absentee landowner. Whatever your status — resident or absentee — it has no bearing on you personally as a forest landowner, but it can have a direct effect on the management of your land.

For example, living on the property will allow you to observe activities and changes on a daily basis. You are likely to possess tools and machinery and be able to perform simple maintenance tasks like mending fences, posting no trespassing signs, and keeping an eye out for insect and disease infestations. An absentee landowner may need to hire someone to carry out these tasks, but an absentee landowner can still be actively involved in the management activities; it just may be more costly because of time and travel expenses. Seasonal activities like tree planting, hunting season, or spring camping trips are wonderful opportunities to spend time on your property and accomplish chores that you want to be involved with. Whatever the case, the absentee landowner is more likely to require assistance to oversee significant forestry operations like harvesting, road construction, installation of wildlife food plots or ponds, and tree planting. Your landowner status can therefore play a significant role in the objective you pursue for your forestland.

Finances of Ownership

There are costs associated with owning forestland beyond the purchase price. Your financial solvency assumes a major role in the decisions you make concerning the management of your forestland. If you decide timber production is a primary objective, be aware there are costs that will have to be carried for some years before you realize a return on your investment. Such costs include reforestation (seedlings and labor), application of herbicide and fertilizer, pre-commercial thinning (you pay to have non-merchantable timber removed), and road construction and maintenance. Consider these questions when evaluating your financial situation:

- What is your time horizon before realizing financial returns?
- Can you afford the upfront costs associated with certain land management practices?
- What other objectives do you want to pursue, and what are their potential costs and returns (e.g. recreation, hunting leases, ecotourism)?

Keep in mind, if timber production is one of your objectives and mature timber is growing on your land, you may be able to realize an immediate positive cash flow if the markets are good. If you decide to pursue this type of activity, make allowances in your budget to cover the costs of establishing a new forest. As a steward of forestland, it is important to get your land back into production as quickly as possible. It is also a good idea to put some of the income from timber sales in an investment account to generate additional capital to cover expenses that may arise.

Family Situation

Your family situation may affect your selection of objectives and management of your forestland. Older forest landowners whose families have moved away have different concerns than those with young growing families. Owners without children or immediate family to inherit their property face different decisions altogether. Because of the long-term character of forest landownership, the benefits your immediate heirs and their children receive are determined by your actions today.

However, your heirs may wish to pursue different objectives for the land. If this is the case, as the current landowner, consider these differences in attitude and gauge the results your actions will have on the future of your family and the land. After all, it is your land and you have control.

Tax Status

Clearly the decisions you make concerning the management of your forestland today will affect the character of your forest in the future. But these decisions will also affect your tax status in ways that you might not realize. Tax matters, especially as they determine when and how to market timber products, should be analyzed by professionals who have experience dealing with forestland. Therefore, consult with a professional forester, an estate planner, and an accountant familiar with forest taxation regulations early in the planning stages. They can guide you in making the best choices to protect your investment.

Knowing what you want from your land is the most important decision you have to make as a forest landowner. Identifying your objectives and assessing personal issues such as family and tax status are key elements in the decision making process. This manual cannot begin to cover issues such as taxes and estate planning, but you must at least consider these matters.

Summary

Well-defined objectives and implementation of prescribed management activities increase the likelihood that you and your family will realize the benefits sometime in the near future. People look at forests differently depending on past experiences and individual desires. Some people simply see the natural beauty of the forest, a dynamic ecosystem working to transform light and nutrients into life. Others view the forest as primarily a place where certain game animals live. Foresters may look at the forest and begin to assess its value as sawtimber or its ability to provide clean water to a stream.

Recreation enthusiasts might see only the potential for hiking, camping trips, trails for horses, or ATVs.

Certain objectives of forest management may not appeal to you personally, but knowing and understanding all the possibilities allows you to make informed decisions. Flexibility in your plan will allow for objectives that may change over time, providing you with the best opportunity for success. Everybody views a forest differently, but there are some objective criteria and characteristics than can be measured. These characteristics will be considered and will lay the groundwork for discussion of the components of a management plan, the focus of Chapter Two.

Review Questions

1. Why is it important for you to evaluate your personal and financial situation at the same time you are selecting your objectives?

2. What distinguishes a resident from an absentee landowner?

3. What are Best Management Practices and why are they important to forestry and society?

4. Financial objectives assume major importance in decision making concerning forest management. How can it affect your management decisions and what are your options?

Suggested Resources

Readings

- Coder, Kim. 1996. "Doing Nothing Costs You! The Blight of Unmanaged Forest Stands." *Forest Resources Unit Publication FOR 96-009.* University of Georgia, Cooperative Extension Service, Athens, GA.
- Korhnak, Larry V. and Mary L. Duryea. 1998. "Forest Resource Information Available on the Internet." School of Forest Resources and Conservation, Florida Cooperative Extension Service, Institute of Food and Agricultural Sciences, University of Florida. Gainesville, FL. *www.sfrc.ufl.edu/extension/ffws/home.htm*

- Varnedoe, Jr. L. (Ed.). 1998. "Forestry on a Budget." The University of Georgia, Warnell School of Forest Resources. *www.forestry.uga.edu/h/publicservice/h/publicservice/publications* (On the Search line, type in Forestry on a Budget.)

Web Sites[1]

- www.sref.info/courses/mtf1
  View the contents of the 2001 Master Tree Farmer (Part 1). This site has text and video for you to learn about the basics of forest management.
- www.forestryusa.com
  One of the foremost Web sites on forests and forestry in America. It provides quick access to all the Internet sites of the federal and state governments, the forest industries, service and supply companies, associations and Non Governmental Organizations (NGO), consultants, education and research, forestry news, employment opportunities, and much more.
- www.fs.fed.us/spf/coop/programs/loa/index
  The US Forest Service Southern Region has a great Web site for private forest landowners with information about taxes and estate planning.
- www.sref.info/publications/online_pubs/
  The Southern Region Extension Forest's web site provides a list of useful publications.

[1] *Note: not all web site addresses include a "www" (ie: http://maps.google.com).*

Chapter 2:
Why Have a Management Plan?

Creating a plan to manage your forest may seem at first like a complicated and intimidating task. Although this process can take a lot of thought and effort, the time and money invested in preparing a management plan will be repaid many times over in the increased profits and enjoyment that you will receive from your land. In addition, organizing the various biological, economic and legal aspects of your property provides you with the tools needed to realize your objectives. Your plan will also take into account the existing conditions of your land, the potential for changing and maintaining these conditions, and the likely results of the objectives you propose.

Your plan will generally contain the following elements:

- A statement of objectives and special considerations for your property
- Maps identifying property location, boundaries, natural features, soil types and timber types
- Stand descriptions that include site characteristics and planned activities (hunting, recreation, timber harvesting, etc.) for each stand
- Schedule of recommended management practices on each stand
- Cash flow of the costs and revenues for all planned activities including possible financial analysis
- Detailed statements that describe any specific practices, especially on environmentally sensitive areas
- Information useful for tax purposes and historical records

This chapter examines how you and the natural resource professionals working with you will develop these elements into a functioning planning document. Before examining the planning process and its components, let us look at the professionals you may call upon to assist you in developing your plan.

The Plan Writing Team

Forest management plans are most often written by professionals with a broad knowledge of forestry practices, especially forest growth and marketing strategies. Usually this combination of attributes belongs to a professional forester. Foresters will play an integral role in the development of your plan. In some states Registered Foresters are the only parties allowed by law to offer such advice for a fee. They are specialists in forestry practices and will help to coordinate a group of experts who might include specialists in soil conservation, resource biology (ecology, wildlife biology, or related fields), recreation, aesthetics, water quality or other pertinent fields. This team, along with you, the landowner, will work together to develop a plan specifically for your forestland.

State foresters, who work for a public agency, may provide at no charge a limited set of services, including examining tracts of timber, survival counts, insect and disease consultations and coordinating the development of management plans. They will meet with you on your land and gather appropriate data. In addition to timber management, the plan they develop will often consider wildlife, water quality, recreational opportunities, current laws and legislation, ways to save on property taxes and ways to gain extra income.

Quite often, state foresters will recommend that you obtain the services of a professional forestry consultant to handle timber sales and complicated management problems. They may also arrange to have certain practices carried out by private contractors, such as tree planting, timber harvesting or boundary surveying.

Some forest product companies offer assistance in planning forest management. After the company forester and landowner meet to discuss the landowner's objectives, the plan is written. These plans usually include an inventory of the property, stand and soil maps, harvest schedule and estimated cash flow of forest management activities. The proper way to harvest and regenerate an area is emphasized by discussing Best Management Practices (discussed in detail in a subsequent chapter). Companies will often act as the landowner's agent by helping to find competent contractors to do necessary tasks such as boundary line maintenance, site prep, planting, prescribed burning, etc. In many instances, seedlings needed to regenerate harvested areas are provided at reduced costs. These landowner assistance programs are provided by forestry companies to ensure an adequate supply of high quality trees. In some cases there is a requirement that the company providing you assistance receive a "first right of refusal." That is, that you are required to offer them the opportunity to purchase your timber (even 30 - 40 years down the road) at a mutually agreed upon price before you offer it to others.

Forestry consultants also provide a full range of services from timber appraisals to detailed management plans. Moreover, because consultants generally receive a percentage of timber sale proceeds, they actively seek ways to enhance timber values to obtain the best prices. Fees for these services may vary from a percentage of timber sales to a fixed daily or hourly rate. Occasionally, consultants will work on a retainer basis, with a set fee for the range of management services required throughout a specified time period. For owners having considerable acreage of substantial value, this type of arrangement more than pays for itself. Even for owners with a small amount of land or with average quality timber, forestry consultants can more than pay for their fees by obtaining higher values for harvested timber than you could get if you were to negotiate timber sales with buyers. Forestry consultants also work to ensure the property and any residual timber stand is protected from damage before, during and after a harvest.

It is important to note that if you decide to utilize the services of a forestry consultant, you should establish a clear understanding beforehand of what services will be performed and what will be charged. Variability in the performance standards and practices of forestry consultants warrant careful attention and research to their selection.

Several different options are available to you in determining who will assist you in making your plan. Even if you have the state forest service draw up a basic management plan, it is important that you develop a *strategy* to effectively manage your forestland. Regardless of whom you choose, they should best represent the needs and desires that you have in mind for your land. Call your state Extension forester or county forestry agency office for a list of forestry consultants in your area (see Appendix B for a list of Extension forestry addresses and Appendix C for a list of state forestry agency addresses).

The Planning Process

The process for creating a forest management plan will usually follow these steps:

- Defining your management objectives
- Examining your forest resources to determine any important features
- Considering alternatives for reaching these objectives
- Developing specific prescriptions for each forest stand
- Organizing priorities for the schedule of activities

You must first determine your long-term goals for your forestland: timber production, wildlife habitat, recreation, or some other use. Professional foresters, along with other specialists, if needed, will then make a resource assessment of your property, equipment, facilities, capital, and experience. This is conducted to evaluate the potential for achieving your goals. The assessment may determine, for example, that moisture conditions preclude certain types of recreational facilities, or that soil erosion problems need to be addressed before logging operations can begin. This assessment will also assist you in determining the specific objectives for your land. Although you probably know the general condition of your forest, it is essential to obtain the specific data needed to create your management plan from professional foresters.

The first chapter focused attention on the need for clear objectives. If, at this point, you still don't have a fairly good idea of what your objectives are, you need to consider carefully what you want to accomplish with your woodland before you can develop any specific actions.

Professional foresters also can assist you in clarifying your objectives. Keep in mind that these foresters, whether they work for a forest products company or for a consulting firm, will probably promote the idea of timber management along with other objectives. Timber production is the objective that promises you the highest financial returns for your investment. You need to examine how well this option fits with the other objectives you have considered.

The Planning Document

All forest management plans are not the same. Your plan will be specifically tailored to fit your objectives and your forestland. It may vary from others by the degree of complexity in your forest, the range of your objectives and recommended practices, the availability of time and money, or the group writing your plan.

In general, forests with similar age classes and species require a fairly uncomplicated management plan. This is especially true when timber production is the primary objective as most stands will be of the same timber type and age. More complex management plans are needed when forests contain several different types of timber or when several different management objectives are being considered. No matter how complex your situation or management plan, the following components should be included in your forest management-planning document.

Map and Area Descriptions

A map should be included in every plan. It will be used as a guide for locating the quality, quantity and distribution of the resources on your land. It will also indicate the boundary lines of your property. A map of your woodland should also identify the following features:

- Map scale
- Orientation and north arrow
- Major landmarks and access routes
- Buildings and other man-made structures
- Bodies of water and streams, with direction of flow indicated
- Stand with forest types and acreages delineated
- Roads and trails

In addition to the map, the following should be included:

- Area or stand number. This number will refer to the specific area outlined on the map of your property, and will be referenced in the written plan.

- Approximate acreage. The estimated acreage for each stand will be included. This is usually computed from your woodland map, or aerial photos.
- Forest age and type. A brief description of the characteristics for each stand as identified in the forest inventory will be included. The resource assessment provides the data for these summarizations.

The major feature on maps for forest management purposes is stand types. Each timber stand should be outlined and labeled in some way in order to delineate it from other stands. For example, a 35 year old stand of loblolly might be labeled L-35. Stands with similar characteristics may be labeled the same and combined for management purposes. Areas without forest cover should also be identified on your map. It is also helpful to obtain aerial photos and soil maps of your property. You can usually obtain soil maps from the USDA Natural Resource Conservation Service.

The Resource Assessment

The information found in your plan will be a summarization of the data gathered during an inventory to assess the resource. The specific data gathered are listed here to give you some idea of what the forester examines and considers in summarizing the data used to develop the recommendations for your plan.

Water and Soil

One area of interest is water quality. Soil type, erodibility potential, and general topographic features of your land all determine its site sensitivity and degree to which best management practices (BMPs) need to be applied to protect water quality. Areas that will help to increase the water quality include streamside management zones (SMZs), wetlands, and drainage networks. These areas need to be highlighted and mapped for special care and protection. Potential water quality problems such as stream crossings, agricultural runoff, and off-site discharges should also be identified and recorded.

The soil type, soil depth, and soil profile may be inspected and recorded. These soil characteristics are frequently the main factor in determining the potential timber productivity for your land. They will also largely determine the type of vegetation your land is capable of supporting. Always be aware of the tree types that are best suited to your soil type.

Wildlife

The presence, arrangement, and condition of suitable vegetation to provide food and cover for desired wildlife are the primary considerations in managing for wildlife. Special care must be taken to enhance your land if wildlife management is one of your principal goals. This is especially true in dealing with any threatened or endangered species. Many laws and regulations strictly protect all endangered species, and special considerations must be taken in your plan not to threaten them or their habitat. Your local wildlife agency biologist can tell you how to manage any protected species found on your land (see Appendix D for state wildlife/natural resource contacts).

Timber

The professional forester will also "cruise" each timber stand in your forest and use the results as a basis for timber management recommendations. A preliminary cruise will identify the tree species and record estimates of the number of trees per acre, the basal area, the size classes, the growth rate, as well as the presence of any diseases or pests. If timber production is your primary management goal, for each stand, data are recorded regarding the site index, merchantable volume, timber quality, and the potential products, such as pulpwood, poles, and sawtimber. Private consulting foresters and industrial landowner assistance foresters are best equipped to handle a timber cruise, as they are experienced in timber inventory and are familiar with local markets.

Recreation

An assessment for forest recreation might also be included in your management plan. Many of the considerations used in other management goals, such as wildlife, vegetation, water resources, and water quality, can determine the recreation possibilities for your woodland. These factors help to define the most desirable locations for recreational facilities like nature trails, cabins, wildlife observation stands, fishing sites, and campsites.

Aesthetics

The appearance of your forest may be the most important consideration that you want to emphasize in your plan. This may be especially true if you live on or near your woodland. Some forestry practices have a drastic effect on the aesthetics of your forest and you may want to plan for some temporary buffer areas when you harvest timber.

All of the information gathered during the resource assessment can be condensed into a brief description for each stand and stated in your planning document. Once completed, this assessment serves as the basis for developing specific management strategies.

Recommended Management Practices

This is the heart of the planning document as it provides a working schedule for you to follow. This section states the specific actions to be taken on each stand. It also gives general dates for when these operations should be performed (e.g. thin stand #6 in the next 5–7 years).

Environment Considerations

Any areas of environmental concern should be identified in your planning document. These might be endangered species, wetlands or erodible trails. These areas might be subject to regulation and protection and will undoubtedly affect several of your recommended actions and operations. Does your woodland contain any environmentally sensitive areas?

Financial Analysis

In many cases, a financial analysis of the management plan is included in the planning document. This analysis requires determinations of present timber values, estimates of timber growth and yield, calculations of future timber values, and projections of future operational costs. Even the best and most current analyses are only estimates, and are constantly subject to change. A financial analysis is only as good as the information available at that time. Nevertheless, such estimates are invaluable for planning and usually worth the cost to develop. Certainly not every landowner needs this information, but the higher the value of your resource and the larger your woodland, the more important this information will be to your decision making process.

Updating and Record Keeping

The time period covered by your management plan will vary according to the complexity and specific objectives involved. Your plan will probably cover at least a five-year period. Longer-term plans need to be regularly evaluated to reflect changes in conditions, markets, and objectives. You may need to make radical alterations in the prescribed activities and operations for specific stands, or for your entire tract. Generally, the longer the planning period covered by your document, the greater the likelihood that changes and adjustments will be needed.

Every activity performed and expenses incurred in your forest should be documented with complete records. These records, besides being important for management decisions, are essential for tax purposes.

Summary

This chapter has dealt primarily with the process of developing a forest management plan specifically for your woodland. The suggestions serve only as basic guidelines and do not take into account your individual situation. You have to decide on an approach in developing a plan that best suits the goals for your woodland.

Above all else, the plan created for you should be consistent with your objectives. This assumes that your objectives are in compliance with ecologically sound forestry and environmentally safe practices. All professional foresters are expected to follow a code of ethics in all aspects of their work. This means that they will work with you to develop sound, attainable objectives that will benefit both you and your forestland.

Review Questions

1. What types of professionals routinely prepare management plans for private landowners?

2. What elements generally are included in forest management plans?

3. Why is flexibility an important attribute of management plans?

4. What environmental concerns do you need to consider when developing a management plan for your woodland?

5. Why is it important to maintain accurate records of all forestry operations?

6. Are your management objectives realistic expectations for your woodland?

Suggested Activities

1. Make an appointment to visit your local county forester. Ask her or him to show you an example of a management plan that might be similar to one you would have prepared. Talk with him about management objectives and forestland opportunities/markets in your area.
2. Prepare a rough management plan for the property you own or manage. Although much of the plan will need to be prepared by a professional, you can fill in important components such as goals, objectives, legal/historical context and other items to begin answering the questions below.
3. Work with your state agency or a private industry/consulting forester to develop your management plan.

Suggested Resources

Readings

- Konrad, Gary C., Catherine A. Albers. 1992. "A Consumer's Guide to Consulting Foresters." *Publication WON-6.* North Carolina Cooperative Extension Service, North Carolina State University. Raleigh, NC.
- Megalos, Mark A. 1995. "Management by Objectives: Successful Forest Planning." *Publication WON-32.* North Carolina Cooperative Extension Service, North Carolina State University. Raleigh, NC.
- Monaghan, Dr. Thomas A. 1992. "Forest Management Alternatives for Private Landowners." *Publication 1337.* Mississippi Cooperative Extension Service, Mississippi State University. Starkville, MS.

Videos

- Kessler, G. and R. Hamilton, 2004. "Master Tree Farmer Satellite Broadcast Videotape (Tape #1)—Forestry Terms and Concepts." Full tape series available to order at www.mastertreefarmer.net. 2001 version available for viewing online at www.sref.info.

Web Sites

- www.sref.info
  Southern Regional Extension Forestry Homepage with links to state forestry extension units, state forestry agencies, state forestry associations and other relevant forest management information.
- www.forestryindex.net
  An online forestry publications library categorized by major and minor topics.
- www.pfmt.org/planning/default.htm
  Auburn University hosts this site relating to private forest management planning.
- www.sfrc.ufl.edu/Extension/ffws/mp.htm
  University of Florida website page on the importance and characteristics of a forest management plan.
- www.state.ma.us/dem/programs/forestry/docs/sampleplan.doc
  An example of a forest stewardship management plan.

Unit II:
Wildlife and Other Forest Uses

Many forest owners derive pleasure from their woodlands by enjoying non-timber benefits such as wildlife and recreation. Although financial returns brought by future harvests or eventual land resale are often long-term goals, leisure activities reward landowners who take the time for a wide range of activities in their woodlands. These activities may be family oriented, personal or commercial. They may be relatively inexpensive to achieve or they may require substantial landowner inputs. No matter whether you reside on your wooded property or visit it only occasionally, you can enjoy continual benefits from your forest's non-timber products.

The size of your ownership creates few restrictions on the amount of enjoyment you can obtain from non-timber uses. However, the size of your woodlands, along with several other factors, definitely affects the types of leisure activities that may be considered economically viable on your property. Formal recreation facilities and huntable populations of most game species require considerable acreage to be successful. On the other hand, bird watching and casual walks for aesthetic beauty can take place on any size tract.

Forests can produce quality timber while offering you tremendous opportunities to enjoy leisure activities. In fact, many practices recommended for improving timber production will improve wildlife habitat and opportunities for recreation. The art of establishing a forest management plan that includes timber and non-timber objectives depends on realizing the inherent limitations of your forest's resources and recognizing the necessary tradeoffs needed to obtain the highest quality for all your primary objectives. When leisure activities emerge as primary objectives in addition to timber production, they increase the complexity of forest management planning.

This unit examines the issues to be considered when management for wildlife and other forest uses are desired. It includes a chapter on managing for wildlife and a chapter on outdoor recreation and aesthetic enhancement compatible with forest management.

Chapter 3:
Managing for Wildlife

Your woodland is the habitat and home for a diverse population of wildlife species. Your actions can dramatically affect these creatures, either to their benefit or their detriment. Likewise, these same actions can enhance or diminish other recreational opportunities for your woodlands. Real estate development, highways and other intensive land use changes have caused a decline in wildlife habitat. Keep these things in mind as you make choices in developing a management plan for your woodlands. As you read through this chapter, consider a few questions that will allow you to make informed decisions concerning your woodlands:

- Why should you manage for wildlife?
- How will certain activities affect the wildlife inhabiting your land?
- How can you coordinate wildlife and timber practices in your woodland management plan?

Objectives and Incentives

Chapter One described the process by which you determine the overall management objectives for your woodland. Once you have defined your primary forest management goals and determined your forest's condition, you are prepared to consider particular wildlife objectives and incorporate them into the overall planning process. Your objectives regarding wildlife may incorporate one or more of the following goals:

- Promoting the greatest variety of species and number of wildlife
- Having one or a combination of game species, such as deer, squirrels, turkey, or grouse on your land
- Promoting only game species
- Encouraging all wildlife for non-hunting recreational such as bird-watching and photography
- Deriving income from hunting, fishing, trapping or other recreational uses and leases

Stewardship and Personal Satisfaction

Being a good steward of the forest resources provides most landowners with a great deal of personal satisfaction. If you enjoy hunting, fishing or just looking at wildlife and knowing that different species are present on the land, forestry practices can then be tailored to improve wildlife habitat and at the same time provide revenue from timber and income from access fees for wildlife recreation.

Privately Owned Forests

Two-thirds of the land in the Southeast is forested and most of this forestland is privately owned. Therefore, habitat and access to most of our wildlife species is in the hands of private landowners like you. Although wildlife is a publicly owned resource, private landowners ultimately determine whether quality habitat will be available to support wildlife populations.

Access and Leasing Privileges

People who use the wildlife resource, particularly those who hunt and fish, are willing to pay for access privileges. Income potential to you comes in the form of hunting leases to individuals or groups or with daily access fees for wildlife recreation. Leases for hunting may provide substantial financial benefits but the lease amount may vary according to local demand and the wildlife species present on your land.

Limitations in Planning

The general condition of your woodlands will dictate how the land should be managed. Several factors should be considered prior to the finalization of your wildlife management plan:

- The natural capabilities of your land to produce and sustain wildlife
- Wildlife species already present on your land
- Habitats provided by adjacent properties
- Wildlife species present on your woodlands and in the surrounding area

The capabilities of the land and the wildlife species present define the limits of the management possibilities. Depending on the size of your woodland, the habitat conditions that are found on neighboring properties may be very important. Managing for large species such as deer or turkey requires large areas of habitat (at least several hundred acres). These species may travel over large areas to meet their habitat requirements and their range may include areas beyond your property. Therefore, the habitat on your neighbor's woodland can directly affect the success of your wildlife objectives.

The abundance and type of habitat present on your property determine the wildlife species best suited for your woodlands. For example, gray squirrels and wild turkeys are dependent predominately on hardwoods and the mast they produce. (Mast is the fruit and seeds of trees and shrubs that provide food for wildlife, e.g. blueberries, acorns.)

Management for these species can be much more difficult if your land is lacking the necessary hardwood tree species. The habitat requirements of various wildlife species must be considered as you evaluate your management possibilities. See Table 3.1 for these requirements.

Timber harvesting opens the forest stand, allowing sunlight to penetrate the canopy to the forest floor, stimulating the growth of forbs, shrubs and young trees important to wildlife. Wildlife species such as deer, quail, wild turkeys and rabbits derive much of their food and shelter from low growing plants and benefit from timber harvesting. Deer benefit from the increased browse and thickets created by timber cutting, conditions that may not have been in abundance in a closed canopy forest. Quail and wild turkeys feed on the newly abundant fruits, seeds and insects. Rabbits find more protective cover and forage in the denser ground vegetation. But in the same respects, timber harvesting can prove harmful to some of these same forest wildlife species as well as others. For example, wild turkeys and squirrels, which require mast-producing hardwoods, will suffer if all the hardwood trees are removed. This is a typical example of a trade-off. You may want to leave some mature trees during timber harvest for the benefit of certain wildlife.

Many timber management practices prove beneficial to wildlife. Often timber management practices can be modified to accommodate various wildlife species. For example, heavier thinning of stands to allow more sunlight to reach the forest floor will benefit wildlife, but will reduce timber production. Longer timber rotations favor wildlife species such as turkeys, which prefer mature timber stands, but will defer income from timber sales.

Costs and Benefits

Costs of Managing Wildlife

Direct costs of wildlife management are normally minor and depend on the intensity of the practice implemented. Planting of food and cover plots, pruning or release cutting of fruit and mast-bearing trees and putting up nest boxes are examples of direct costs. But most of the cost of managing forests for wildlife will be indirect, in the form of reduced income from timber because of tradeoffs. This cost can be slight to significant, but in most cases will not be excessive. Indirect costs can be calculated by forecasting timber production under various management options with and without wildlife habitat enhancement activities.

Species	Spring	Summer	Fall	Winter
Deer	Browse Grasses Forbs Escape cover	Browse Grasses Forbs Escape cover	Browse Acorns Escape cover Wintering cover	Browse Acorns Escape cover Wintering cover
Turkey	Insects Nesting cover Escape cover Grasses Forbs	Insects Soft mast and berries Brooding cover Escape cover Grasses Forbs	Acorns Roosting cover Soft mast Escape cover Grass and weed seeds	Acorns Roosting cover Soft mast Escape cover Wintering cover
Quail	Insects Nesting cover Escape cover Grass and weed seeds	Insects Brooding cover Escape cover Grass and weed seeds	Roosting cover Escape cover Wintering cover Seeds from trees, shrubs, vines and herbaceous plants	Roosting cover Escape cover Wintering cover Seeds from trees, shrubs, vines and herbaceous plants
Grouse	Insects Nesting cover Fruits and drupes Buds and flowers	Insects Brooding cover Escape cover Fruits and drupes	Escape cover Wintering cover Grass and weed seeds Buds Grapes	Escape cover Wintering cover Grass and weed seeds Soft mast Evergreen forbs and buds
Squirrel	Den sites Buds and flowers Hard mast	Leaf nest sites Buds and flowers Berries Fruit	Den sites Hard mast Soft mast	Den sites Hard mast Soft mast

Table 3.1: Seasonal requirements of some selected wildlife species.

Benefits

The benefits derived from wildlife management can be produced in several forms:

- Recreation for you, your family and friends
- Income from the sale or lease of hunting and fishing privileges
- The personal satisfaction derived from abundant wildlife

Income derived from providing hunting and fishing privileges to the public is normally obtained by either selling daily permits to individuals for a fee or leasing to an individual or group for a seasonal or annual fee. The going rates for permit and lease fees vary according to location, game species available and the abundance of those species. Advice on appropriate fees can be obtained from local landowners who presently lease their land for wildlife recreation, wildlife officials, or your local county extension agent.

General Habitat Requirements for Wildlife

All wildlife species require the same basic necessities of food, shelter, water and living space. The kinds and amounts of these requirements differ among species, each having its own particular preferences. These preferences include specific foods and types of cover, as well as adequate home range areas to obtain all of their habitat needs. The larger animals, such as deer and wild turkey, require relatively large areas. Wild turkeys, for instance, require at least 700 acres; whereas smaller animals such as squirrels and quail require less territory. If big game is your interest, but your woodlands are too small for proper management, you might explore combining several adjoining ownerships as a management unit.

The single most important aspect of management in providing wildlife habitat is achieving as much diversity as possible. In general, the more variety your forest has in tree and plant species composition, size and age classes in conjunction with cover thickness and edge between cover types, the better the wildlife habitat will be. ("Edge" is the distinguishable line where one cover type ends and another begins.)

Game Species Habitat Requirements

Gray and Fox Squirrels

Hardwoods are preferred by gray squirrels, whereas fox squirrels prefer mature stands of cone-producing pine trees. Gray squirrels require partial hardwood stands of trees old enough to produce mast (25 years) and provide dens (35 years). Their home range is 2 to 8 acres. Gray squirrels need about 1.5 pounds of hard mast per week from September through March. Principal late fall and winter hard mast foods of squirrels are nuts, particularly hickory nuts, pecans, walnuts, beechnuts and acorns. Stored nuts, berries, soft mast, buds, seeds, fungi and insect larvae are supportive foods for squirrels in spring and in periods of poor hard mast production. Managing for gray squirrels on your property requires retaining some hardwoods. Existing food producers and den trees should be retained as well to guarantee adequate shelter and sustenance.

Bobwhite Quail

Quail like open areas, (a reduced basal area down to 65 square feet/acre or less), mature pine forests that are burned with controlled fire every 1-2 years, newly cutover tracts, young pine plantations (1-4 years old), old-fields and agricultural habitats. Quail may also do well in other forest types and simply require forests interspersed with openings, brush and grass.

Edges between forest types or between the forest and open clearings generate an assortment of grasses and brush preferred for nesting. A nesting pair of quail requires a minimum clearing of 1/5 acre; they generally nest from April to September. Approximately 85 percent of an adult quail's diet consists of seeds. Legumes, grass and various weed seeds are the most important food sources and are preferred in the order listed in Table 3.2. Summer provides an increased supply of grasshoppers and other insects important to quail, especially young chicks whose diets are comprised almost entirely of insects. The normal home range for quail is 40 acres.

Age Class	Spring	Summer	Fall	Winter
0–7	Browse Grasses Forbs Annual weeds Berries Nesting cover Insects	Browse Grasses Forbs Annual weeds Soft mast (grapes) Berries Brooding cover Insects	Browse Grasses Forbs Grass and weed seed Annual weeds Soft mast (grapes) Berries Roosting cover Escape cover Wintering cover	Browse Grass and weed seed Roosting cover Escape cover Wintering cover
8–20	Buds and flowers Limited browse	Leaf nest sites Soft mast Limited browse	Leaf nest sites Soft mast Limited browse	Leaf nest sites Soft mast Limited browse
21+	Buds and flowers Den sites Nesting sites	Leaf nest sites	Hard mast Soft mast Den sites Roosting sites for turkey Limited wintering and escape cover	Hard mast Soft mast Den sites Roosting sites for turkey Limited wintering and escape cover

Table 3.2: Habitat characteristics of natural unmanaged stands.

White-Tailed Deer

Deer are extremely adaptive creatures and survive in almost all forest types (and increasingly so in suburban areas). The early stages of timber rotation and intermediate cuts produce abundant deer browse and fruits. Prescribed burning and fertilization attract deer because of improved nutrition and palatability of food plants.

During fall and winter, deer prefer hard mast (acorns, beechnuts, pecans, etc.) and evergreen forage. The spring and summer months provide rapidly growing green browse and herbaceous plants as principal foods. Deer require about six to eight pounds of green food daily for each 100 pounds of body weight. Generally, a whitetail will spend its entire life within one mile of where it was born. Its home range seldom exceeds 300 acres if food, cover and water are adequately interspersed. Deer seek bedding spots that reflect a need to keep out of the wind, to keep warm or cool, to keep dry or perhaps to avoid insects. Understory and underbrush development on your woodlands may be an essential element to whitetail management.

Ruffed Grouse

This game bird is found in the southern Appalachian Mountains and the Cumberland and Appalachian Plateaus, usually above 2,000 feet in elevation. It prospers in the early stages of forest succession but occurs in mature stands as well.

Grouse use fruit, seed, catkins, buds and green parts of over 300 plants for food. Broods require insects from late May through July. Thickets, vine tangles and dense shrub growth are used for escape cover. Understories near openings in the forest and old logging roads serve as brood range and nesting cover. Grouse require 40 to 50 acres for their home range.

Wild Turkey

Good turkey habitat contains mature stands of mixed hardwoods, groups of conifers, relatively open understories, scattered clearings, well-distributed sources of water and reasonable freedom from disturbance.

Frequent and sustained disturbance by free-running dogs, vehicles or people may cause turkeys to temporarily forsake portions of their range. A turkey's range is about one square mile (640 acres). Openings and clearings throughout the home range are essential for brood range. Favorite sites for nesting are under the slash from tree tops. Turkey diets consist primarily of grass and seeds of herbaceous plants in the fall, mast and forage in winter and spring, and forage and insects in summer. Acorns, dogwood berries, clover and pine seed are important foods. Fields of soybeans, corn, chufas and small grains as well as pasture are frequently used by turkey.

Rabbit

Rabbits prefer small openings in the forest and brushy areas with low-growing vegetation. Cottontails are associated with old fields, cutover areas, young pine plantations and mature, control-burned pine forests; swamp rabbits inhabit bottomland hardwood forests and other wetland habitats. Honeysuckle provides important food and cover for rabbits. Clover and grasses are preferred foods. Brush piles and fallen trees provide escape cover.

Dove

Dove are found in range and agricultural habitats, but they do nest and roost in heavily wooded areas such as open pine forests, pine plantations and brushy hardwoods. Dirt roads are utilized by dove for dusting and resting. Dove are seed-eaters and feed on the ground in open areas. Dove consume millet, wheat, other small grains and a variety of weed seeds. Pokeberry is a preferred food, as are paspalums, panicums and crotons. Management for dove may require large areas of non-timberland or newly cutover timber areas.

Waterfowl

Ducks, particularly wood ducks and mallards, are found in wooded areas with bodies of water. Ponds and swamps for nesting, resting and feeding are necessary. Wood ducks nest in hollow trees or in man-made nest boxes. Large beaver ponds make excellent waterfowl habitat.

Further Detailed Information

The game species listed above are common animals of interest in terms of wildlife management. Of course many other game species exist in the Southeast, and all of them can be properly managed. For additional information refer to the literature in the suggested readings section at the end of this chapter and contact your state wildlife agency or state forestry commission office to obtain further information.

Habitat Requirements for Non-Game Wildlife

A variety of non-game animals found in the forest can be managed by using various wildlife management practices. These species are sometimes the most interesting inhabitants to some woodland owners. All animals play some role in forest health. Woodpeckers help control damaging forest insect pests, while hawks and owls control small rodent populations. Many species of songbirds control insect populations and contribute to forest diversity. But the most important non-game wildlife you should consider is endangered species. If endangered species are present they *must* be incorporated in your management plans; in many cases they dictate what management actions can be practiced.

Hawks and owls require snags and large hollow cavity trees. Owls nest in hollow trees, and both use snags for perching while hunting. Furbearing animals, such as raccoons, mink, muskrats and beavers, require streams, ponds or other wetland areas as habitat. Muskrats and beavers do not range far from water, but raccoons and mink may range over upland areas even though they are primarily dependent on wetlands. Beavers require bottomland habitat, which can be flooded by building low dams, which they construct from brush, timber and mud. They also need woody vegetation on which to feed. Their flooding and feeding habits may cause damage to your timber in some locations, but these habits also provide excellent wetland habitat for waterfowl and other furbearers, as well as slowing storm runoff and storing water. Trapping can control beavers, but you should consider both the benefits and the possible damage when planning how to manage these animals.

The kind and number of songbirds in your woodlands will vary with both the season and the kinds of cover present. Some species prefer deep, more mature woodlands, while other species use early succession stages.

Endangered Species

Eagles

Habitat requirements for endangered species depend on the particular species. Eagles (bald and golden), although they have been returned to safe population levels, are still considered threatened and are given special protection by law. Eagles inhabit open areas and wetlands, where they may find prey species such as fish, rabbits and other small mammals, and/or waterfowl.

Large, dead trees, called snags, are important for perching, and live trees are preferred for nesting. Eagles are migratory birds, seen most often on the coast or in the mountains, and are not likely to be found on most small woodlands in the interior of the Southeast.

Red-Cockaded Woodpecker

The red-cockaded woodpecker is a resident species in some mature pine stands and is listed as an endangered species. These birds live in colonies and must have suitable cavity trees and foraging areas to survive. Cavity trees are conspicuous because of the flow of sap down the trunk and the fact that they are usually flat-topped. Typical red-cockaded woodpecker habitat consists of fairly open stands of mature pines (80+ years). Management of the birds includes saving the occupied nesting and foraging areas, which may consist of 50 to 100 acres, and keeping the understory open by use of controlled burning. This can represent a considerable cost (and loss of timber production) to a landowner.

Every year more species are added to the endangered species list. Before conducting any forest management practices on your land, you must find out if any endangered species reside in your woodlands. Information concerning endangered species and any other species of interest is available from your local conservation office or the U.S. Fish and Wildlife Service *(www.fws.gov).*

Recommended Management Practices

These forestry practices may be beneficial to wildlife (both game and non-game):

- Breaking timber management units into small blocks (100 acres or less if possible), to achieve a variety of age and size classes. Even-aged timber management in small units of different age, with well-distributed regeneration and intermediate cuts, provide interspersion of stands and productive understory conditions for wildlife.
- Making clearcuts irregularly shaped to achieve maximum edge. Avoid large clearcuts (50 acres or more) if possible. Deer, for instance, use the outside edges of a clearcut much more heavily than large open areas.
- Leaving hardwoods along drains and streams to provide travel lanes, escape cover and food for deer, turkeys, squirrels, and other wildlife. These streamside management zones (SMZs) also protect streams from siltation and clogging with logging debris. Avoid damage to both hardwood and pine wetlands while logging.
- When thinning or practicing timber stand improvements, leave snags and den trees standing (at least 2 to 3 per acre). Remove only those which are suppressing timber growth or present safety hazards during logging operations.
- Leaving key areas of large diameter mast-producing hardwoods scattered through the area. These should be at least 1/4 acre and larger if possible.
- Seeding roads and other small openings to grasses, clovers or other game food and cover plants adapted to your area. This provides food for an assortment of wildlife and brood openings for quail and turkeys. Also, cut trees to open the canopy along roads, allowing light to penetrate and enhance edge growth. These areas may also be allowed to grow up in native grasses and legumes. They can be maintained as open space by periodic disking, mowing or controlled burning.
- Using controlled burning on pine sites in the coastal plain and Piedmont to stimulate growth of legumes and other wildlife food plants. Controlled burning should usually be done in small blocks in a 3 to 5 year rotation, beginning when pine stands are pole-sized and following thinning operations. Burn in winter or early spring (December, January and February are usually the best months).

Additional practices can enhance wildlife habitat:

- Disking between rows of pines or in open stands to stimulate the growth of grasses and legumes important to wildlife. (Do not plant bicolor; it's invasive!)
- Allowing windrows (raked piles of slash after logging) to grow up with volunteer vegetation for several years in newly cut timber stands.
- Planting small openings (1/5 to 1/2 acre) to grass or clover, with a natural shrub border 10 feet wide to provide brood cover for quail and turkeys and winter feeding areas for quail, turkey and deer. Where deer are the primary species to be managed, openings can be larger (1 to 5 acres). There should be at least one opening per 100 acres.

- In agricultural fields adjacent to woodlands leaving unharvested border strips of crops such as corn and soybeans to provide food for game.
- Erecting squirrel nest boxes in or adjacent to hardwoods where den trees are not present in sufficient number. Wood duck nesting boxes should be placed in ponds or other water areas. Beaver ponds can be managed to provide excellent waterfowl habitat by controlling the water level. Drain the pond in the summer, allowing for native wetland plants to grow or planting exposed mud flats in Japanese millet, and reflood it in 60 to 80 days.

Available Resources

Numerous agencies are available to provide you with advice and assistance. See the appendices at the end of this book for a listing of several state and federal agencies. State wildlife agency biologists can provide on-the-ground advice for both habitat and use of the wildlife resources. Wildlife enforcement officers can provide assistance with law enforcement and control of hunting. State forestry agencies provide assistance with development of timber, wildlife management plans, the use of controlled fire and other silvicultural practices. Natural resource consultants (consulting foresters and wildlife biologists) can also provide technical assistance for a fee for developing and implementing wildlife management plans.

The Natural Resources Conservation Service can assist you with farm and woodland plans, including soils information and management advice. The Cooperative Extension Service can provide general advice on forest and wildlife management and direct you to the proper person for more detailed information and assistance.

Wildlife food and cover planting materials for wildlife management can be obtained free of charge from state wildlife agencies in some states. Other food and cover planting materials can be obtained from commercial nurseries. Check with your local Cooperative Extension Service Agent for such information. Plans for building nest boxes for squirrels and wood ducks can be obtained from extension or state wildlife offices.

Review Questions

1. What three factors determine wildlife management possibilities in a woodland?

2. Name three potential costs and three potential benefits of managing wildlife.

3. Give four reasons why it is important to leave hardwood drains and streamside management zones.

4. How is controlled burning beneficial to wildlife?

5. Name three agencies that provide advice and assistance on wildlife in woodlands.

Suggested Activities

1. Defining management goals and objectives is the first step in developing a wildlife habitat management plan and helps provide a "road map" for planning and implementing improvement practices. Take time to reflect on what your management goals and objectives are and write them out in a bulleted format. Incorporate these goals and objectives into your current forest management plans or plans that are to be developed for wildlife.
2. Decide what wildlife species you are interested featuring on your land. Take time to understand the life history, biology and habitat needs of these species and management practices that might help improve habitat for these wildlife on your land.
3. Become familiar with the native wildlife food and cover plants that occur on your land. Make every effort to mark and protect these areas before any forestry practices occur.

Suggested Resources

Readings

- Baskett, T. S., M. W. Sayre, R. E. Tomlinson and R. E. Mirarchi. 1993. "Ecology and Management of the Mourning Dove." A Wildlife Management Institute Book. Stackpole Books. Harrisburg, PA.
- Bellrose, F.C. and D. J. Holm. 1994. "Ecology and Management of the Wood Duck." Stackpole Books. Harrisburg, PA.
- Bidwell, T.G., S.R. Tully, A.D. Peoples, and R.E. Masters. 1992. "Habitat Appraisal Guide for Bobwhite Quail." Circular E-904. Oklahoma State Cooperative Extension Service, Oklahoma State University. Stillwater, OK.
- Byford, James L. and Thomas K. Hill 1995. "Fish and Wildlife Conservation." *Publication 972.* Tennessee Agricultural Extension Service, University of Tennessee. Knoxville, TN.

- Byrd, Nathan A. 1985. "A Forester's Guide to Observing Wildlife Use of Forest Habitat in the South." *Forestry Report RB-FR5.* USDA Forest Service.
- Byrd, Nathan A. and Herman L. Holbrook. 1974. "Forest Management Bulletin: How to Improve Forest Game Habitat." USDA Forest Service.
- Dickson, J. G. editor. 1992. "The Wild Turkey: Biology and Management." A National Wild Turkey Federation Book. Stackpole Books. Harrisburg, PA.
- Halls, L. K., R. E. McCabe and L. R. Jahn. editors. 1984. "White-tailed Deer: Ecology and Management." Wildlife Management Institute. Stackpole Books. Harrisburg, PA.
- Hamilton, Rick A. 1984. "A Guide to Information About Forest and Wildlife Management." *Publication WON-2.* North Carolina Cooperative Extension Service, North Carolina State University. Raleigh, NC.
- Hayes, Don, Rebecca Richards, and Edwin J. Jones. 1994. "Wildlife and Prescribed Burning: Fire as a Forest Habitat Management Tool." *Publication AG-457.* North Carolina Cooperative Extension Service, North Carolina State University. Raleigh, NC.
- Hamel, P. B. 1992. "The Land Manager's Guide to Birds of the South." The Nature Conservancy, Southeastern Region. Chapel Hill, NC.
- Hazel, Robert B. and Edwin J. Jones. 1992. "Deer Management." *Publication WON-12.* North Carolina Cooperative Extension Service, North Carolina State University. Raleigh, NC.
- Jackson, J. J. 1980. "Bring Ducks to Your Land." Georgia Cooperative Extension Service, University of Georgia, Athens, GA.
- Jackson, J. J., G. D. Walker, R. L. Shell, and D. Heighes. 1981. "Managing Timber and Wildlife in the Southern Piedmont." University of Georgia, Cooperative Extension Service Bulletin, Athens, GA.
- Jones, Edwin J., Peter T. Bromley, Mark A. Megalos, and Rick A. Hamilton. 1995. "Wildlife and Forest Stewardship." *Publication WON-27.* North Carolina Cooperative Extension Service, North Carolina State University. Raleigh, NC.
- Miller, H. G. and K. V. Miller. 1999. "Forest Plants of the Southeast and Their Wildlife Uses." Southern Weed Science Society.
- Miller, K. V. and L. M. Marchinton. Editors. 1995. "Quality Whitetails: Why and How of Quality Deer Management." Stackpole Books. Harrisburg, PA.
- Payne, N. F. and F. C. Bryant. 1994. "Techniques for Wildlife Habitat Management on Uplands." MacMillan Publishing Company, New York, NY.
- Payne, N. F. 1992. "Techniques for Wildlife Habitat Management of Wetlands." McGraw-Hill. New York, NY.
- Rosene, W. 1969. "The Bobwhite Quail: Its Life and Management." Rutgers University Press. New Brunswick, NJ.
- Traugott, Timothy A. 1996. "Forestry/Wildlife Myths and Misconceptions." *Publication 1612.* Mississippi State Cooperative Extension Service, Mississippi State University. Starkville, MS.

- USDA Forest Service Handbook. 1971. "Wildlife Habitat Management Handbook. Southern Region." U.S. Department of Agriculture. FSH 2609.
- Wood, G. W. 1988. "The Southern Fox Squirrel." *Brookgreen Journal* Vol. XVIII, No.3.
- Yarrow, G. K. and D. T. Yarrow. 1999. "Managing Wildlife: Managing Wildlife on Private Lands in Alabama and the Southeast." Sweet Water Press.

Web Sites

- www.whmi.nrcs.usda.gov
  Wildlife Habitat Management Institute website. Online publications on wildlife habitat management.

Videos

- "Deer Management Research." Cooperative Extension Service—Louisiana State University. Tape # V084. Key concepts are given in the area of deer herd management and food plot management for different seasons of the year.
- "Get in the Game! The National Wild Turkey Federation's Guide to Attracting Wildlife to Your Land." (CD). The National Wild Turkey Federation *(www.nwtf.org)*. Edgefield, SC.
- "Master Tree Farmer: A Shortcourse for Private Landowners." Video taped series from satellite course available from the Web sites *(http://mastertreefarmer.org/)* or *(www.sref.info/)*.
- "Master Wildlifer: A Shortcourse for Private Landowners." Video taped series from satellite course available from the Web sites *(http://masterwildlifer.org/)* or *(www.sref.info/)*.
- "Quality Deer Management: Proven Techniques for Managing Your Deer Herd." Part I of the Quality Deer Management Association Video Series. AAI/Our Gang Productions. Nashville, AR.
- "Quality Deer Management: Enhancing Habitat on Your Hunting Lands." Part II of the Quality Deer Management Association Video Series. AAI/Our Gang Productions. Nashville, AR.
- "White-tailed Deer Management" (video). Alabama Cooperative Extension Service, Auburn University, Auburn, AL.

Chapter 4:

Nature-Based Tourism Opportunities for Your Woodland

Nature-based tourism is one of the fastest growing industries in the world. The appeal to get back in touch with nature has become stronger and stronger, especially for those who live in urbanized, high pressure, and high stress environments. You can take advantage of this renewed interest by developing recreational facilities on your woodland. Qualities of your forest, like wildlife, its history, ecological processes, and ancient artifacts, are interesting to many Americans and visitors to America, and bird watching is currently the fastest growing outdoor recreation activity in the world. As such, you may be able to develop and potentially profit from such outdoor activities as bird watching, hiking, camping, hunting, fishing, picnicking or studying nature.

The use and management of your natural resources for outdoor recreational activities depends upon several important factors. This chapter will explore factors to consider before becoming involved in nature-based tourism and tools to implement it.

Things to Consider Before Taking the Leap

People Management Skills

One important consideration in determining the potential for a recreation business is the skill and desire to work with the public. No matter how well your land lends itself to recreational development or how much you learn about the business, you need to possess good people skills if your recreation venture is going to be successful.

In almost every type of recreation business, you will be working with people of all types in many different situations. Catering to the needs of these customers can be trying and demanding. As a result, your interest in the public should be given serious consideration before you make any investments in recreation for your land.

How Will you Benefit?

Nature-based tourism is popular because it can provide a diverse array of benefits to people and the environment. Identifying and clarifying the benefits you are specifically interested in is key to ensuring you meet your future management goals. For example, if you want to make a lot of money, then you might be thinking how to get many people on to your land and what facilities are needed to handle large crowds. However, if money is only a small benefit you desire, and you are looking at nature-based tourism as a way to help conserve your land or "share your story," you should think small and work to control the number and type of visitors to your property.

Financial Planning

Nature-based tourism must eventually be profitable in order for a landowner to invest in the enterprise. Anyone who is starting a new business understands it takes financial resources to invest in new facilities and supplies (i.e., capital investments) and regular expenses to operate that business (i.e., operating expenses); the larger the business, the greater the costs. At the least, the landowner must conduct a thorough cost benefit analysis before investing the money into a potential nature-based tourism operation. This will require a significant amount of detective work. In other words, what are those hidden costs, which might derail the entire operation? For example, restrooms might need to be built or people must be hired to take reservations. These types of costs are high and should be paid for by the revenue generated from visitors.

Land Characteristics

The characteristics of your property will determine, to a large extent, the feasibility of establishing a recreational business. Before you invite people to your land know exactly where you want them to go and what you want them to see. What makes your forest special? Do you own a wetland that attracts a diversity of birds and wildlife? Do you have cultural resources, like Native American mounds or an old cabin, where visitors can learn more about the area's history? Do you have different forest types that help tell a story about how your family has cared for the land? Are trails already established on your property?

Something on your property needs to serve as the opportunity to give people unique experiences. In rare cases, the resource might attract visitors without your working to make it attractive (e.g., an aesthetic waterfall, wetlands that attract birds, or historical artifacts that people can visit). However, in most cases, you will need to offer facilities and/or services to provide opportunities for visitors to attain unique experiences. These might include guided tours through your forest, interpretative signs, or mountain bikes available for people to rent.

You will need to look at your property in a different way to identify potential attractions. In fact, you might want to invite a local tourism professional to tour your property and help you see your resources through potential visitors' eyes.

Recreation Options for your Woodland

Choosing the right type of recreation activity for your land will depend primarily on the location and physical features of your land. Areas containing a diverse natural resource base are generally capable of supporting several types of recreational activities. Numerous options for recreation are possible on your timberland. Let us look at some of these.

Hunting

Leasing the hunting rights of your land to individuals or clubs is the most common way forest landowners earn money from recreation in the Southeastern United States. Payments for hunting leases are generally made on a per acre basis for access to your property, usually for a specified period of time. Leases may be for all game species or for one species. For instance, if you have a good deer or turkey population, consider leasing to two parties instead of one.

In addition to the previously mentioned factors that affect recreational decisions, the choice to provide hunting on your property will also depend upon the size and diversity of your land, the wildlife resources that are found on your land, and the time and length of hunting seasons in your area. These considerations will help you determine the management method that is best suited to you and your land. These options include day-use permits, day-use with room and board, and seasonal leases.

Hiking

Creating hiking trails and walkways is one of the most common and reasonable forest recreational developments. These trails can be used as either a part of the recreational experience itself or as a means to reach other recreational areas. Careful attention needs to be taken in their design so as to easily facilitate travel and reduce erosion while preserving and protecting plants, animals, and scenery along these trails.

Hiking has not been considered to be a source of revenue in the past. Recently, however, the demand for less crowded and "undiscovered" areas has sharply increased so as to make this a profitable recreational activity. Again, your land needs to contain some distinctive features, such as a unique scenic or historic site, in order to justify charging a fee to hike on your land.

Fishing and Water Recreation

If you are fortunate enough to own property that contains lakes, rivers, or streams, you may be able to create extra income by developing some type of water-based recreational venture. The most obvious choice is fishing. Revenues may be derived not only from fishing fees, but also from selling bait, refreshments, fishing equipment, or from renting boats. It may be a good idea to get assistance from your local Soil Conservation Service and local fish and wildlife agency to develop your lakes or streams for fishing. Some localities may classify these kinds of activities as retail business and require zoning for business.

Other water recreational options may include swimming, canoeing, picnicking, and boating. These types of activities are especially attractive to families, for safe, day outings. Developed and marketed correctly, these types of facilities can become very profitable.

Pulling It Off

Identify your Visitors and Market your Attractions

Although nature-based tourists come from around the world, the safest place to begin marketing your nature-based tourism operation is close to home. Focusing on visitors in your local community or region is cheaper and more efficient than taking a much larger marketing approach (e.g., interstate billboards). As your operation becomes more successful and you can afford to broaden your market, you can begin to pursue more expensive and far-reaching advertising strategies.

To begin an economically sustainable business requires marketing research. Unless you have a surplus of money that can be used to conduct this marketing analysis you should rely on your knowledge of your community and local tourism professionals to help you identify the type of people your property might serve. Chambers of commerce, tourist development councils, and visitor and convention bureaus welcome new businesses and look forward to helping your business succeed. In fact, tourist development councils are supported by county tourism development taxes, which might be available to you to aid in marketing and promotion.

Liability Concerns

Generally, on primitive, undeveloped lands, liability is minimal. However, as the recreational site is developed and fees are charged, you begin to assume more responsibility. How much of a financial and liability risk are you willing to assume? Two strategies are available to assist in controlling risk — liability insurance and business structure.

First, recreation providers should consider obtaining special liability insurance to help protect against lawsuits. By budgeting for insurance premiums, which are a fixed cost, as opposed to a potential unknown cost that could amount to a staggering loss, you will be protecting your business and assets.

Second, various business structures differ from each other primarily in three ways:

- The way the business is taxed

- The way capital can be raised
- The amount of liability that the owner is responsible for

Generally, you will want to develop a business structure to protect personal assets from your business assets. Types of business structures include sole proprietorship, general partnership, joint venture, limited venture, limited partnership, limited liability partnership (LLP), limited liability company (LLC), corporations ("C" and "S"). Often limited liability operations are the most efficient business structures for new operations. More information can be obtained by talking to your county's small business development office or chamber of commerce, and especially to your accountant and attorney.

Access

Recreational facilities should be easily accessible and within reasonable driving distance of large population areas to ensure a nearby market for your business. Factors such as proximity to undesirable businesses, dense populations, or high traffic areas could be deterrents to the creation of a positive recreational experience and must also be considered. Also, visitors need to be able to find your area. Marketing and promoting will be discussed later, but clear signs along the highways and rural roads must be established so visitors can find you.

Scenery Management

Maintaining the scenic quality of your woodland is especially important in providing recreational facilities to the general public. Many people associate clear-cuts, controlled burns, or herbicide applications with damaged or blemished forests, thus reducing the recreation potential. Such negative impressions can be avoided by establishing sound silvicultural practices. With proper planning and layout, all thinning, harvesting, prescribed burning, and reforestation activities can be conducted in a manner that will increase the visual appeal of your forests. For example, consider the following options when harvesting timber on your land.

- Adjust removal operations near the forest edge to create a park-like transition, from dense forest to open spaces or fields.
- Favor a mixture of hardwood and pine species for variety, especially near stand edges.
- Alter harvest selection criteria on forest edges. Retain residual trees of varied form, shape, blossom or autumn color, and wildlife value instead of selecting trees solely for greatest timber or fiber value.
- Establish irregular, outlying clumps of trees to create a natural appearance of the forest edge.

Visitor Management

Nature-based tourism means inviting people onto your property, but you are the one in control of what visitors will do when they are there. It takes work to ensure the people come and have a quality recreation experience. The following suggestions are key to promoting and managing visitors.

Visitors to natural areas want to learn. Research shows that learning is often one of their main motivations for going to a forest or other natural area. As a landowner, you want your visitors to be educated. Not only will they understand something of great value to you – your land and home, but they will also learn how to treat it respectfully.

Control access to your property. In the early stages of a tourism business, you could take reservations and have potential visitors schedule specific times to visit your property. This may be easiest with local groups who do not have to travel far; otherwise, setting up regularly scheduled tours (e.g., every Wednesday between 1:00 and 5:00) would be appropriate. If you plan to allow hunting on your land, a hunting lease is the best way to control access. A clear, concise contract should be drawn up so all concerned parties are fully aware of their rights and responsibilities. Consult your attorney and the local Cooperative Extension Agent to obtain more detailed information on hunting liability and hunting leases and permits.

Partnerships

Nature-based tourism requires you to reach out to other businesses and organizations. Visitors have a diversity of needs. They need transportation, places to eat and sleep, and other recreation areas to visit. You cannot serve all their needs; therefore, building partnerships with other organizations will provide improved experiences for your visitors, ensure that tourists find their way to your property, and spread the benefits of nature-based tourism out to other members of your community. Potential business partners include:

- Restaurants
- Trail clubs
- Hotels, motels, bed and breakfasts
- Horse owners association
- Antique stores
- Photography clubs
- Souvenir stands
- College geology classes
- Grocery stores
- State forestry associations
- Rental car agencies
- Catering businesses

Volunteers are also becoming a heavily used resource for nature-based tourism operations. Some community members might enjoy spending time working in your forest and talking to visitors. Before taking on volunteers, a landowner must realize that volunteer labor is not free. There will be some cost either in time or labor to train and supervise volunteers. In fact, many public land management agencies hire coordinators simply to manage their volunteers.

Summary

Providing recreation opportunities on private lands is similar to providing other forest resources—it requires much planning and management. It also requires you to look at your land in a new and different way. But, when managed effectively, outdoor recreation opportunities can provide valuable benefits to you, your family, and community. It might also place a high value on resources (e.g., a wetland or historic building), which were not valued under traditional forest uses. You likely know that you manage important and valuable resources. Using these resources as recreation opportunities might be your opportunity to show them off and make a little money in return.

Review Questions

1. How would you define nature-based tourism?

2. To develop nature-based opportunities what kinds of features are needed on your land? What kind of personal skills are needed?

3. Describe how each of these professionals might be needed—attorney, accountant, county planning office, marketing specialist, building contractor.

4. Partnerships may be the key to a successful operation. On your land, with your specific tourism opportunities, who might be your partners?

Unit III:

Establishing a Forest

Many landowners face the decision of establishing a new forest at some point during their ownership tenure. Reforesting a cutover woodland, converting poorly stocked or low quality stands to more productive forests, or retiring marginal cropland are typical situations that you may encounter. Upon completing Unit Three, you should understand and be able to analyze these reforestation opportunities on your forestland.

Landowners should analyze the productive capacity of their soil before investing in a long-term forestry investment. The old adage "plant your sorry, worn out acres to trees" is not always wise advice. Timber, like any other crop, is most productive on fertile sites, and severely eroded or depleted soils may not be suitable for timber production—or at least a good rate of return on your invested dollars.

Chapter Five investigates soil factors that affect tree species selection, timber yields, and economic return. Chapters Six and Seven discuss the necessary steps to achieve successful pine and hardwood regeneration. Because harvesting the existing stand is often the first step in reforestation, these chapters begin with harvesting strategies and then discuss the alternative methods of site preparation as well as natural versus artificial methods of seeding or planting.

The bottom line in any investment is economic return. Forestry is blessed with numerous financial incentives including long-term capital gains, exclusion of income, reforestation tax credits and deductions, preferential property tax treatment, and cost-share funding available from state and federal sources. Chapter Eight discusses these incentives and then analyzes the reforestation investment under several assumptions.

Chapter 5: Forest Soils and Site Index

Soil quality is a very important factor that should be addressed before undertaking any forest management activity. Soils will determine which trees species yield the greatest timber volume and when to harvest. Ultimately, knowing the soil quality will help you decide what level of investment to make to yield an acceptable economic return from your forest management.

Soils vary greatly in their ability to produce merchantable volumes of pulpwood, sawtimber, veneer, poles, piling or other wood products in a reasonable period of time. Soils also determine suitability for campsites, trails, ponds, and other multiple uses. Soil factors have important implications in timber harvesting and erosion control as well. Landowners must be aware of soil factors that affect forest production before investing in forest regeneration or management. For these reasons, Chapter Five examines important soil characteristics and the method used to identify site quality.

Soil Factors

The following factors have a major impact on forest soil productivity and site index.

Topsoil Depth

The depth of the uppermost soil layer is a critical factor affecting tree growth. Topsoil is highest in organic matter and nutrients, is usually well aerated and drained, and allows maximum root growth and root penetration.

Soil Texture

The proportion of sand, silt and clay in the topsoil and subsoil layers is called texture. Sandy soils are normally very well drained and often lack nutrients due to constant leaching loss. At the other end of the spectrum are the pure clay soils made up of very small, fine soil particles. Clay soils are often very hard and poorly drained, characteristics that can impede tree root growth. Soils with a mixture of sand, silt and clay provide for the best tree growth.

Subsoil Consistence Class

Consistency of the subsoil layer (the layer directly underlying the topsoil) is another important factor in forest soil productivity. The combination of soil-sized particles and the physical and chemical properties of each individual particle type in a given soil determines the soil's consistence class. Consistence ranges from very friable to very plastic.

- very friable: coarse textured; very loose when dry.
- friable: cannot be molded without crumbling.
- semi-plastic: tends to facture when molded.
- plastic: can be molded continuously and permanently.
- very plastic: cannot be broken with fingers when dry

Limiting Layers

A layer which restricts the downward penetration of a tree's root system will reduce tree growth in direct relation to the depth of the layer. In rare instance a limiting layer may increase site productivity, such as on sandy soils where the layer may retard leaching of nutrients and increase available moisture.

Fertility

Trees grow over a wide range of soil fertility levels. Fertilization is normally not recommended early in the rotation except in the case of a critical deficiency of a major nutrient such as phosphorus. A soil test prior to site preparation will alert a landowner to critical deficiencies. Research has shown conflicting results in forest tree response to nitrogen fertilization, particularly early in the rotation. Tree growth may actually be suppressed if the fertilizer increases the growth of competing weeds. Best results from early fertilizer use arise in combination with herbicide or mechanical control of competing vegetation. Late rotation fertilization done 5 to 8 years before final harvest increases timber yields in many situation, but may not be economically practical.

Internal Drainage

Few tree species can grow in constantly wet soils. Drainage can be improved or the impacts of poor drainage minimized in some cases by tiling, ditching or adding bedding as a site preparation method, but these, too, may not be economically feasible practices.

Site Index (SI)

The collective influence of soil factors will determine the site index for a particular tree species on a given soil area. Site index is the total height to which dominant trees (the tallest) of a given species will grow on a given location. For loblolly pine, a "SI (age 50)=70" tells us that we can expect loblolly seedlings planted on that area to be 70 feet tall when the trees are 50 years old. Index age and tree species must be stated because the site index will differ between species growing on the same area. A close relationship exists between site index and timber yield. Both value and volume of merchantable wood improve with increases in site index (Table 5.1).

Measuring Site Index

Site index can be determined by two methods. One method is to find, on the area in question, several dominant trees of a species for which a site index curve has been developed. Using accurate age and height measurements, one can read the site index from a graph showing height over age

SI (age 50)	Stumpage Yield (at Age 40[1])	Value (at age 40)
70	6 MBF[2] + 26 cords[3]	$1,850[4]
80	14 MBF + 37 cords	$3,725
90	19 MBF + 38 cords	$4,750
100	29 MBF + 37 cords	$6,725

[1]Stumpage yield at age 40 is defined as the volume of timber standing (on the stump) at age 40.
[2]Thousand board feet (MBF) of sawtimber per acre assuming recommended thinnings are completed.A board foot is equal to a 1-inch by 12-inch board that is 1 foot long. For this example, volume was calculated using the International Rule. The International Rule is one of three common log rules, or formulas used to estimate volume or product yield from logs and trees.
[3]Cord of pulpwood per acre removed as thinnings.
[4]Value per acre assuming $200 per MBF and $25 per cord.

Source: North Carolina Cooperative Extension Service Woodland Owner Note No. 7, "Forest Soils and Site Index."

Table 5.1. Relation of Site Index (SI) to Timber Yield of Managed Loblolly Pine

curves for that species (Figures 5.1, 5.2, 5.3, 5.4, 5.5, 5.6). For example, in Figure 5.1, checking a 20-year old stand with 60 ft. high trees that are considered dominant would give you a site index of 70 for that stand.

The second method of determining site index is based on physical characteristics of the soil. Tables giving site index by this system have been published for several important species. Generally, foresters will know about tables developed for conditions in their working area (see Tables 5.2 and 5.3 as examples). Necessary information about the soil includes depth of the topsoil and plasticity of the subsoil. In deep sands, the depth of a finer-textured horizon and fine particle content of that horizon are used instead of topsoil depth and subsoil plasticity.

Site index can be calculated with reasonable accuracy for virtually any commercial tree species. A landowner can consult a professional forester to evaluate the site indices for the tree species on a particular property. Site index information is included in the comprehensive soil surveys completed within the last few years or in progress in many counties. The county Natural Resources Conservation Service can provide information related to soil surveys.

Some landowners may ask the question as to whether site index can change naturally over time. As is the case with natural systems, this indeed can happen, but the time frame is often hundreds of years.

Incorporating Soil Management into Your Management Plan

A management plan for your property allows you to determine your personal objectives, manage efficiently, avoid costly errors, make knowledgeable decisions and evaluate your progress. Soil characteristics found on your land directly influence the type of vegetation that will be present. Describe the soils found on your property by including in your management plan the soil series name, characteristics and relation to forest management. Details on soil properties that influence tree

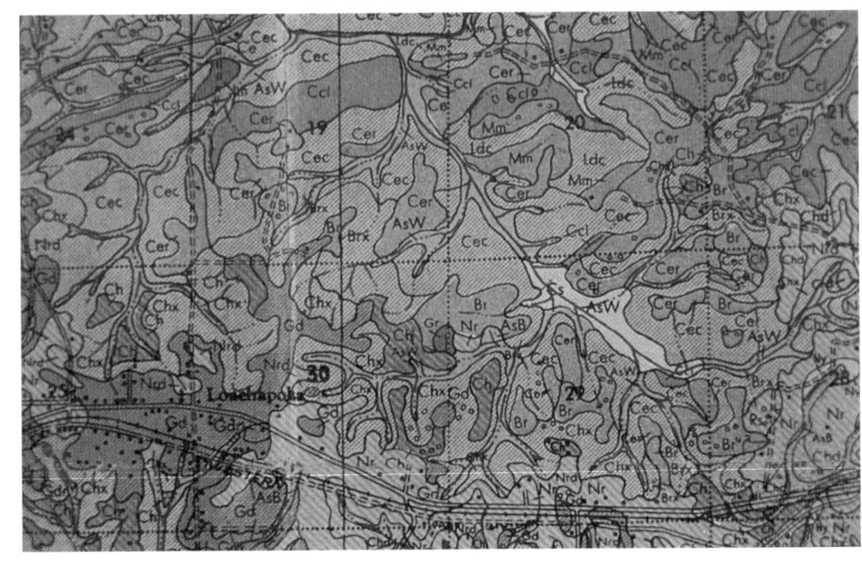

Figure 5.7. "A Soil Map can be a Valuable Tool in Developing a Management Plan for Your Forestland."

Credit. John Beck

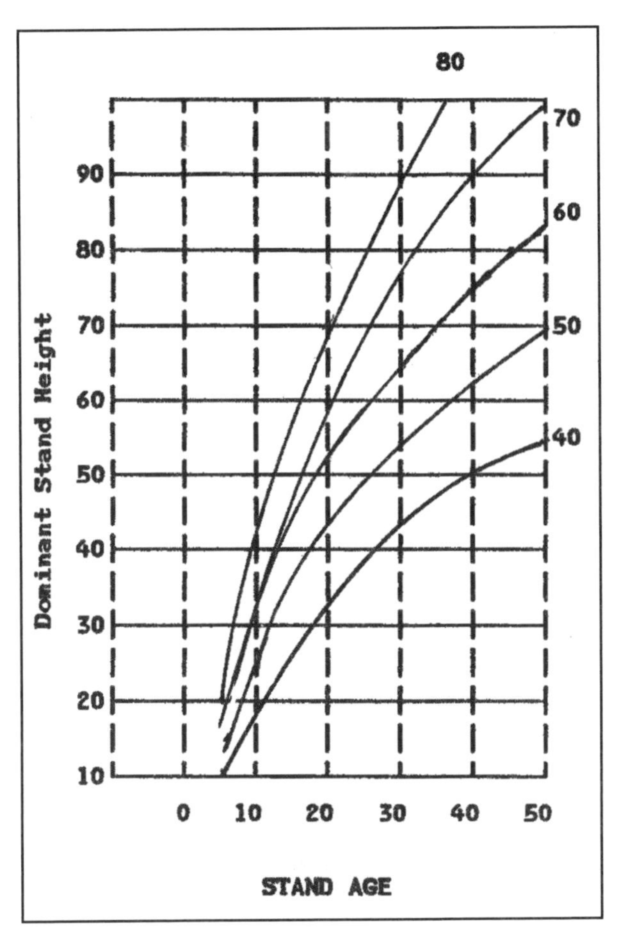

Figure 5.1. Site Index Curve, Loblolly Pine, Base Age 25 Years

Credit: North Carolina Division of Forest Resources, "Forester's Field Handbook," 1988. (Titled, "Smalley & Brower, 'Loblolly Pine Plantations, Site Index Base Age 25 Years, Piedmont, Old Field'")

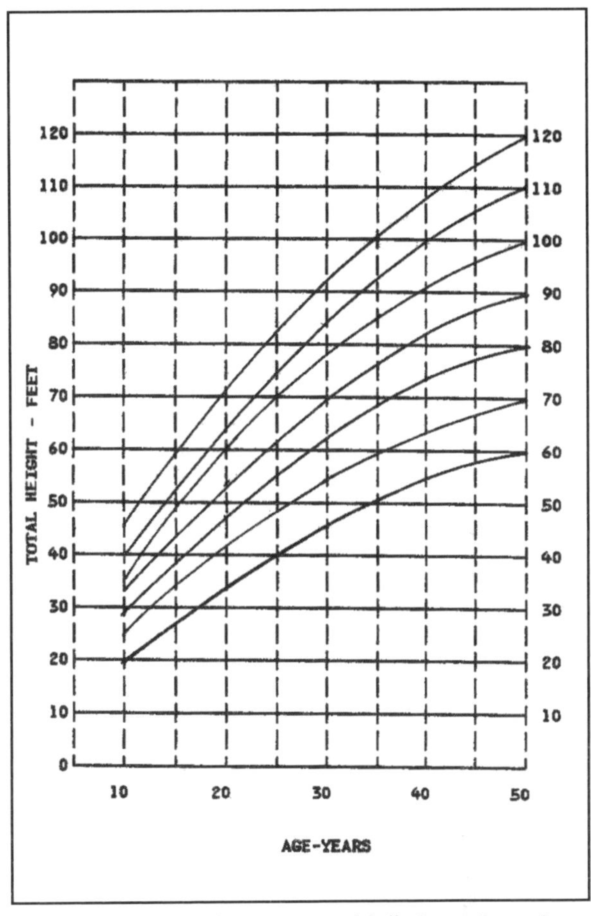

Figure 5.2. Site Index Curve, Loblolly Pine, Base Age 50 Years

Credit: North Carolina Division of Forest Resources, "Forester's Field Handbook," 1988. (Titled, "Coile & Schumacher, Jour. For. 51, 'Adjusted site index curves for young loblolly pine stand. Piedmont Natural & Plantation stands, Base Age 50'")

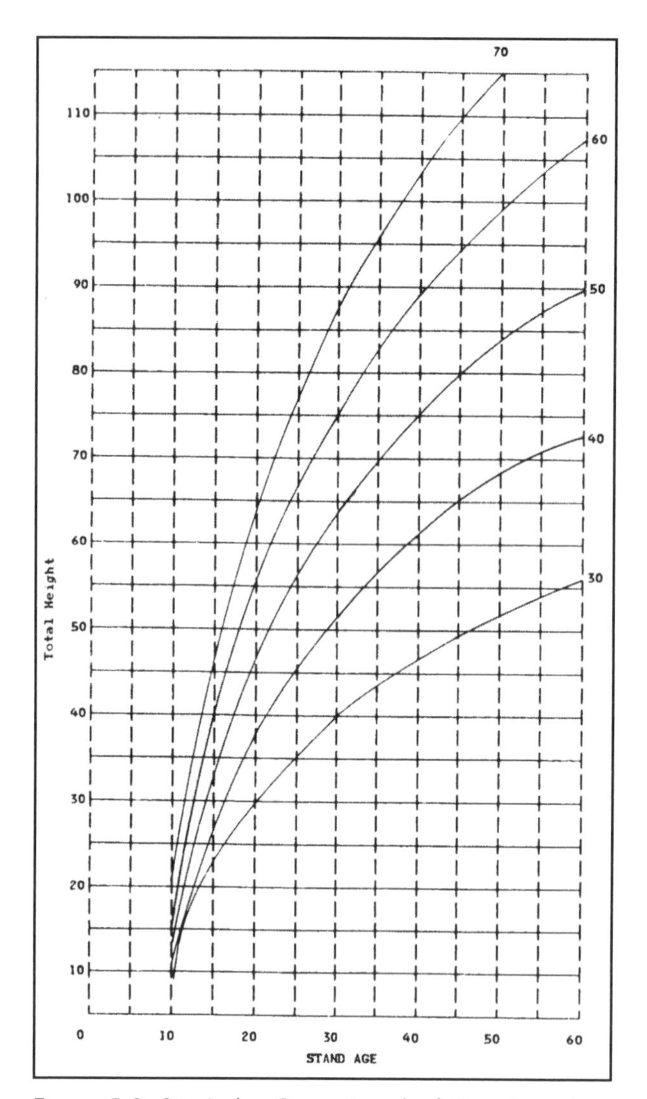

Figure 5.3. Site Index Curve, Longleaf Pine, Base Age 25 Years

Credit: North Carolina Division of Forest Resources, "Forester's Field Handbook," 1988. (Titled, "Farrar's Jour. For. 71, 696, 'Longleaf Pine, Site Index Base Age 25, Plantation'")

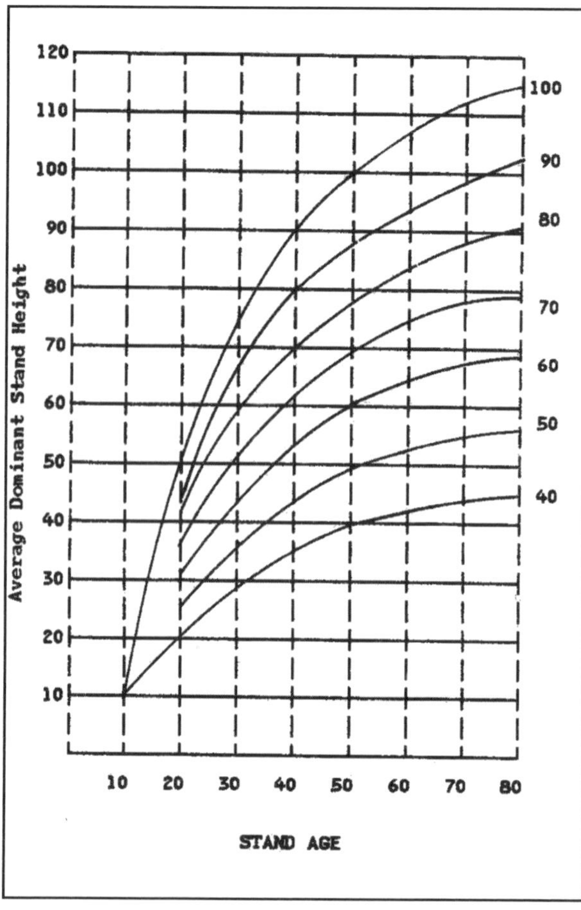

Figure 5.4. Site index curve, Longleaf Pine, Base Age 50 Years

Credit: North Carolina Division of Forest Resources, "Forester's Field Handbook," 1988. (Titled, "Longleaf Pine, Longleaf Pine, Site Index Base Age 50, Southeast Natural Stands")

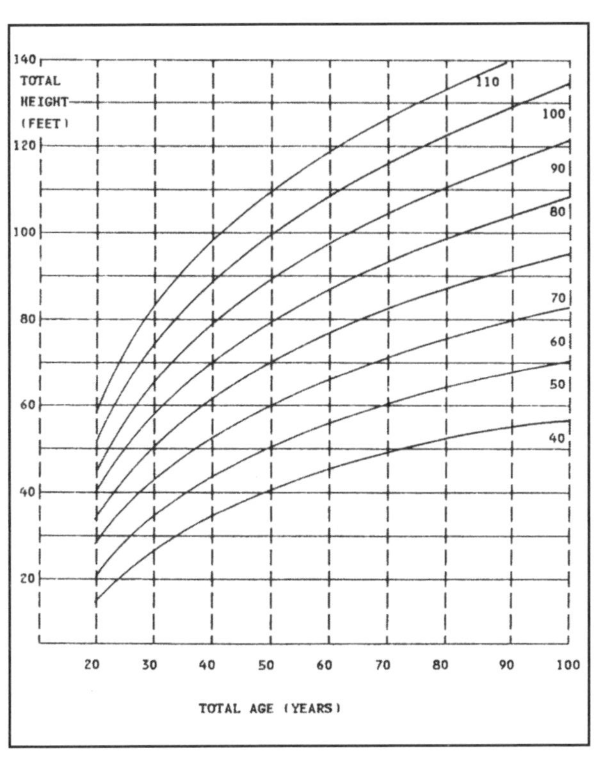

Figure 5.5. Site Index Curve, Red Oak, Base Age 50 Years

Credit: North Carolina Division of Forest Resources, "Forester's Field Handbook," 1988. (Titled, "SOURCE: Doolittle, W.T., SOIL SCIENCE, Vol.22, No. 5, Sept - Oct., 1958; Schur, G.L., USDA, Tech. Bul. No. 560 (1937); Gevorkiantz, S.R., Lake States Exp. Sta., Tech. Note No. 485 (1957). 'Red Oaks, Height in Feet of Average Dominant & Codominant Trees, by Site Index at 50 Years, In Eastern U.S.'")

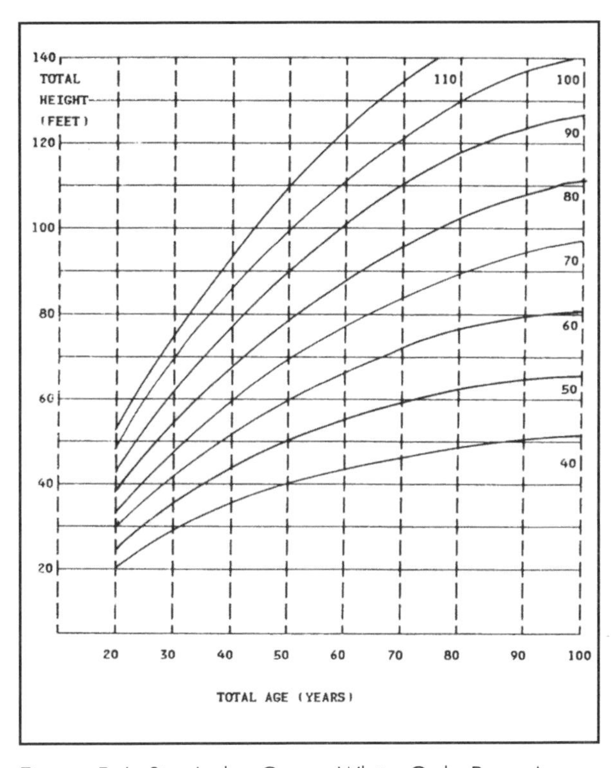

Figure 5.6. Site Index Curve, White Oak, Base Age 50 Years

Credit: North Carolina Division of Forest Resources, "Forester's Field Handbook," 1988. (Titled, "SOURCE: Doolittle, W.T., SOIL SCIENCE, Vol. 22, No. 5, Sept.-Oct., 1958; Schur, G.L., USDA, Tech. Bul. No. 560 (1937); Olsen, JR. D.J., SE. For. Exp. Sta. Res. Notes No. 125, (59). 'White Oaks, Height in Feet of Average Dominant and Codominant Trees, by Site Index at 50 Years, in Eastern U. S.'")

Subsoil Characteristics		**Depth of Topsoil in Inches**					
Consistence when moist	Texture	6	12	18	24	30	36
		Site Index					
Very Friable (noncohesive)	Sands	65	70	73	75	77	79
Friable	Loamy sands to light sandy loams	70	75	79	81	83	85
Friable	Sandy loams	73	79	82	85	87	89
Friable	Loams	75	81	85	88	90	92
Semi-plastic	Sandy clay loams to clay loams	77	83	87	90	92	94
Plastic	Sandy clays	78	85	89	82	94	96

Source: North Carolina Division of Forest Resources, "Forester's Field Handbook," 1988.

Table 5.2. Site Index of Loblolly Pine in the Costal Plains of Virginia, North Carolina, and Northeastern South Carolina as Influenced by the Characteristics of Well and Imperfectly Drained Soils for 50 Year Old Trees

Subsoil Characteristics	**Depth of Topsoil (Inches)**						
Consistence when moist	0–2	4	6	8	10	12	18
	Site Index						
Very friable (noncohesive)	51	62	66	68	69	70	71
Friable	47	59	62	64	65	66	67
Semi-plastic	43	54	58	60	61	62	63
Plastic	38	49	53	55	56	57	58
Very plastic	33	44	48	50	52	52	53

Source: North Carolina Division of Forest Resources, "Forester's Field Handbook", 1988.

Table 5.3 Site Index of Shortleaf Pine in the Piedmont Plateau as Influenced by Soil

growth, like soil depth, texture and productivity, can be located in soil survey reports. Soil maps and reports may have information about the engineering aspects of your soil for logging and road building. A soil map or overlay should be included to go with the property map when several soils are found. Additionally, important information such as soil erosion potential and soil drainage may also be found in the soil survey reports.

An example of a complete soil description in a management plan for a floodplain and upland area follows:

> The floodplain and adjacent slopes in this area contain a variety of soil types including Mecklenburg Sandy Clay Loam, Chewacla Sandy Loam and Wehadkee Silt Loam. The Mecklenburg series is found on the uplands adjacent to the creek. It exhibits slight erosion and windthrow hazard (trees uprooted by excessive wind), while equipment limitations, seedling mortality and plant competition are all moderate. It is an average site for upland hardwoods and yellow pines. This soil series ranges from fair to poor as a producer of wildlife cover and foods.
>
> The Chewacla and Wehadkee soil types are found in the floodplain of the creek. They are moderate to severe in the areas of equipment limitations and plant competition; windthrow hazard is moderate. All of this is due to the wet nature of these soils. Both soil types are good for the production of bottom land hardwoods, especially willow oak and yellow poplar. These soils also make a very high production site for Loblolly Pine. The soils range from poor to good in wildlife plant production with the Chewacla being the better soil, especially for hardwood (i.e. oak) trees and coniferous (i.e. pine) species. The Wehadkee soil is more wet-natured and is considered a good producer of wetland plants.

Species Selection

Selecting the proper tree species to manage or establish on a particular site involves several steps:

- Determine your objectives. If timber production is the major objective, select a species that will economically produce timber products. If wildlife, recreation, aesthetics or other uses are the objective, select species accordingly. On many tracts, several uses may be compatible.
- Select species with a proven track record in terms of growth and acceptance in the local market.
- If two or more species could be selected and timber production is the major objective, select the species that will yield the greatest dollar return in the shortest amount of time. In most cases, this means selecting the species with the highest site index.

Landowners should invest only in those species on those acres capable of producing an acceptable economic return. Returns will vary with investment, species, and site quality. Foresters should provide information including site quality, yield, and projected economic returns before deciding which, if any, species should be managed on a particular property. For more information see the suggested readings below.

Summary

Soil characteristics affect tree species selection, timber yields, suitability for various activities, and economic returns from forest management. Forest owners should evaluate the site index of their property before selecting which tree species, if any, are suited for economical timber management. Species selection for management or reforestation should be based on the landowner's objectives and expectations relative to timber yield and/or from multiple uses such as wildlife and recreation. A management plan can assist forest owners in systematically and efficiently planning and accomplishing objectives.

Suggested Resources

Readings

- Borders, B.E. and R.L. Bailey. 1997. "Loblolly Pine and Pushing the Limits of Growth." Technical Report CAPPS 1997-1. University of Georgia, Athens, GA. *www.forestry.uga.edu/docs/loblolly.html.*
- Brady, N.C., and R.R. Weil. 1996. "The Nature and Properties of Soils," Eleventh Edition. Prentice Hall, Upper Saddle River, NJ. 740 pp.
- Carmean, W.H., J.T. Hahn, and R.D. Jacobs. 1989. "Site index curves for forest tree species in the eastern United States." St. Paul, MD.: USDA Forest Service, North Central Forest Experiment Station.
- Hamilton, R.A. 1997. "Forest Soils and Site Index." Publication WON-7. North Carolina Cooperative Extension Service, North Carolina State University. Raleigh, N.C.
- North Carolina Division of Forest Resources. 1988. "Forester's Field Handbook." Division of Forest Resources, North Carolina Department of Natural Resources and Community Development.
- USDA Forest Service, Southern Research Station. 1995. "Long-term soil productivity research in the South." USDA Forest Service, Southern Region and Southern Research Station.

Review Questions

1. What is site index and how is it calculated?

2. What are soil conditions that provide for the best tree growth?

3. What effect does site index have on timber yield? Timber value?

4. How do you measure site index if you do not currently have trees growing on the site, or if the trees that are there are not the trees you want?

5. Why should soils information be included in your forest management plan?

Suggested Activities

1. Measure the height and age of several dominant trees on your property. Work with your forester to determine the site index based on these measurements.
2. Obtain a soil survey report for your county from the Natural Resources Conservation Service. Locate your property on the aerial photos in the report. Make a list of the soils found on your property, and identify key characteristics like soil texture, drainage, suitability of timber, suitability for wildlife habitat elements, etc.
3. Take a shovel or soil auger to several different locations in your forest. Dig a hole and look at the soil features. Is your soil sandy or clayey? Is it wet or dry? Rocky or gravelly? Is there nice, thick, dark topsoil? Or is the subsoil, sometimes red clay, located at the surface?

Chapter 6:

Southern Pine Reforestation

The pine forests of the South have shown themselves to be one of the most highly resilient forest types of the United States. They have been harvested, cleared, plowed, replanted, high-graded, and "poked at" for nearly 200 years, yet they continue to dominate the Southern landscape and economies. Perhaps the fundamental reason for this continued success is their generous capacity to regenerate themselves. Our southern pines often produce abundant amounts of seed, grow fast, and do well on a variety of soils and sites. They are very effective at finding an open area and getting established. A great many of today's forests were once old farmlands that "seeded in" during the middle of the last century.

Even though southern pine forests are prolific regenerators, they often could use a little help, particularly if the landowner is interested in maximizing financial return. Pines generally require open conditions with abundant light in order to "take hold" and do well. A site that has grown up in hardwoods or is otherwise weedy, may need some help to get a new pine stand re-established. Not only do reforestation areas sometimes need a little help, but non-forested areas will obviously require some management to get a stand of pines established.

A Major Fork in the Skid Trail

There are two major pathways to regenerate southern pine forests. The first is known as "artificial regeneration" which is the planting of seedlings or the sowing of seed on the site. The second is "natural regeneration" which uses trees already on the site as a source of seed or sprouts that will establish the next stand. Both techniques can be successful, but they are very different in their cost, methodology and appearance. The major disadvantage of artificial regeneration is the cost. When you include the cost of site preparation, seedlings, planting, and possibly post-planting care, a landowner can easily spend up to $200 per acre in reforestation costs. Natural regeneration, on the other hand, offers a cheap and highly cost effective alternative to these expenditures as long as the owner is willing to put in the time necessary for proper execution. Just because it is "natural" regeneration does not mean that you stand back and let Mother Nature take its course. There are several points to consider when deciding on going down the different paths of natural or artificial regeneration.

Species and Genetics
Because natural regeneration relies on the existing seed and sprout source, it is not possible to change the species to one that might be more suited to the site. The landowner is also missing out on the genetic improvement that forestry research has produced over the last 40 years.

Potential Seed Tree Loss
Natural regeneration typically requires the owner leave a certain number of high quality trees on the site to serve as seed sources. These trees can be lost to windthrow, disease and insects. Also, there may not be enough volume remaining to interest a logger or timber buyer when it is time to harvest the seed trees.

Variable Stand Regeneration
Sometimes natural regeneration results in an overabundance of young seedlings, often called "dog-hair" stands. Or, the spacing may be uneven, with overstocking in some areas and understocking in others. Certainly there are no tree rows that might facilitate any cultural operations in the future.

Regeneration Delay
Artificial regeneration allows for stand regeneration the year following harvest while natural regeneration is typically a 3 to 5 year process. Some owners may prefer the "get in" and "get-out" approach of planting trees after harvest.

Both the artificial and natural pathways can lead to successful stand establishment. And while natural regeneration may have a number of disadvantages, studies have shown it to produce net present values (return on investment) equal to, or even above that of artificial regeneration. You must carefully weigh the advantages and disadvantages of each approach before choosing your path. Your county forester, consulting forester or industrial landowner assistance forester can give you specific recommendations for your tract.

Preharvest Treatments

Whether choosing the artificial or natural regeneration pathway, you need to understand that the foundation for successful pine regeneration is laid BEFORE the harvest of a standing crop. Landowners should begin to lay the groundwork for successful regeneration through a series of preharvest treatments that will improve the success and cost effectiveness of both artificial and natural regeneration techniques. Pine survive and grow best in the absence of competition, regardless of whether they are planted or "seed" in naturally. There are silvicultural practices that landowners can put into place during the years leading up to the harvest that will help the regeneration phase.

Prescribed Burning

Prescribed burning is one of the most cost effective silvicultural tools available to keep hardwoods and other brushy weeds out of pine stands. When used under proper fuel, moisture and weather conditions, fire is an effective and relatively inexpensive method to control small diameter, thin-barked hardwood species such as maple and sweetgum. Burning also reduces the accumulation of fuel and reduces the risk of wildfire. It may be necessary, depending on density of weed competition and amount of fuel, to treat with more than one burn. A well conducted burning program usually results in a "cleaner" looking stand that could increase harvest income as well as reduce preparation costs for regeneration.

Prescribed burning may be done in loblolly, shortleaf, slash, longleaf, and pond pine stands. These species have a thick well-insulated bark when over 4 inches in diameter and 20 feet in height. Eastern white pine, spruce pine and Virginia pine are thin-barked and will not tolerate prescribed burning. Prescribed fires with the objective of cleaning a stand and reducing the cost of the regeneration phase may begin 6 to 8 years prior to the final timber harvest and be repeated at 2 to 3 year intervals with the final burn done within two years of the harvest. Prescribed burning costs vary by tract size; the smaller the tract the higher the per acre cost. Typical costs range from $10 to $15 per acre. Prescribed burning is not a job for amateurs, and trained, experienced professionals should be used for any prescribed burning program. In addition, each state has specific rules governing the conditions for permitting a burn. Landowners should check with their state forestry organization, extension service, or a consulting forester for legal requirements prior to burning.

Timber Stand Improvement

Timber stand improvement (TSI) is the killing or removal of individual trees of undesirable species, poor form, or those infested with insects and disease prior to the final harvest. Not only will TSI increase the growth of the remaining trees, but it could facilitate regeneration and lower the costs by reducing undesirable seed and sprout sources and eliminating highly competitive individuals left after a commercial harvest. TSI is often carried out in stands where all hardwoods exceed 4 inches in diameter or in stands where prescribed burning is not recommended, such as in the management of white pine or Virginia pine which are not fire resistant. TSI may be done manually, mechanically, or with herbicides. Trees killed may often be used for firewood, posts or other wood products. Depending on the method used and density of weeds, TSI normally costs from $40 to $80 per acre.

Harvesting and Logging

Harvesting is the most rewarding and at the same time most potentially damaging single operation in forest management. Specific and more detailed recommendations regarding logging operations are to be found in other chapters of this course. Landowners must be aware that when conducting harvesting operations, Best Management Practices (BMPs) should be employed to prevent non-point source (NPS) pollution and protect water quality. Most BMPs plan for streamside management zones (SMZs), stream crossings, road location, construction and maintenance, timber harvesting, site preparation, reforestation, etc. Poor logging practices can cause irreversible soil damage resulting in lost soil productivity, polluted waters from non-point sources of pollution, and destruction of wetland habitats. Due to the technical nature of BMPs, landowners should seek assistance from a consulting forester, county extension agent, or state forest organization.

While logging does not *directly* impact southern pine regeneration, it can profoundly affect regeneration success through indirect means. Landowners should take extreme care to avoid logging with heavy equipment during wet periods which can easily lead to soil degradation. Seed or seedling survival and growth may be severely impacted by improper logging that results in soil compaction, puddling, and gully or sheet erosion. Soil health is absolutely essential to any forest regeneration success and sustainable forest management. You should be an active participant in the planning and execution of timber harvests conducted on your land. The importance of landowner participation is doubly critical when natural regeneration will be used to establish the next stand. Natural regeneration depends upon the seed from trees purposely left on the site at the time of harvest. If these trees are damaged, then disease or windthrow may make them unavailable for any future harvest, and they might even die before they can make seed for the new stand. Landowners, in consultation with their professional forestry advisors, need to actively monitor the harvesting operation to ensure that future seed trees are cared for as the valuable assets they are.

Natural Regeneration

Natural Regeneration Fundamentals

There are several natural regeneration systems that may be used to regenerate southern pines. But regardless of the system used, there are four basic components to success:

- A suitable seed bed
- An adequate supply of quality seed
- Open, free-to-grow conditions
- A little cooperation from Mother Nature in the form of adequate rainfall and soil moisture

A successful seed catch, germination, and establishment of southern pines requires adeq
bed preparation. Most forest soils have a top layer of pine straw, leaves, and other undecompos
plant materials. Pine seed do not germinate well in this material and it must be removed or pushe
aside to expose bare mineral soil if natural regeneration is to be successful. If the stand has been burned periodically during the final years of the rotation, the site may already be clean enough for a seed catch. If not, then this litter layer must be burned.

Successful natural regeneration also depends upon the young germinant having adequate light and moisture to grow. If there is a great deal of hardwood or brushy competition, the stand may not get established. Even in the presence of a bare mineral soil and a good seed catch, recently germinated pine seedlings cannot compete against the light and moisture needs of a brushy stand. The preharvest treatments discussed above are critical if a landowner is faced with a pine stand that has thick woody undergrowth.

Seed Tree and Shelterwood Systems

The most common and successful natural regeneration techniques used to naturally regenerate the southern pines are the seed tree and the shelterwood methods. Both rely upon trees left standing on the site at the time of harvesting which will serve as seed trees. Later, after the site has seeded-in and the new stand is established (typically two to five years), the seed trees can be removed.

The seed tree method leaves a recommended number of "leave trees" at the time of harvest. This is sometimes referred to as a "regeneration cut" with the number of trees left based on their diameter and species (Table 6.1). The seed tree system works well for prolific seeders such as loblolly, slash, shortleaf and Virginia pine. "Leave trees" should be as evenly spaced as possible, and selected based on superior health, form, and a vigorous well developed crown. Typically the best trees are left as seed trees. Seed trees should be clearly and plainly marked prior to the regeneration cut, *including a paint spot at the bottom of the trunk,* to minimize accidental harvesting of these superior individuals. Seed normally falls from late summer through November, and therefore the regeneration cut should be completed early enough in a good seed year that there is time to conduct any necessary seed bed preparation. Good seed crops occur every 3 to 5 years and can be predicted a year in advance because it takes two years for cones to mature. Depending upon the success of the seed catch and germination, the seed trees may be removed at two years after seed fall. This is sometimes referred to as the final harvest and should not be delayed too long after seedlings are established in order to minimize seedling damage during the final harvest.

DBH	Slash Loblolly	Shortleaf	Virginia
10	12	20	6
12	9	14	5
14	6	12	4
16+	4	12	4

Table 6.1 Minimum Number of Seed Trees Per Acre

The shelterwood method is similar to a seed tree method except more trees are left on the site during the regeneration cut to serve as seed

	: of Shelterwood Trees
	38–57
	30–45
	21–29

Table 6.2 Number o. ıelterwood Trees Commonly Recommended Per Acre

sources (Table 6.2) and to provide a more sheltered environment within the stand. Not only are there more seed sources, but the added wind and sun protection provided by those trees can increase survival of some species such as white pine. Shelterwood is preferred for nonprolific seed producers, such as longleaf pine, but may be used successfully with all southern pine species. One distinct advantage of the shelterwood method over the seed tree system is that the shade of a shelterwood can decrease weed competition. Moreover, a shelterwood leaves sufficient timber volume per acre to ensure an operable final harvest to remove the shelter trees. A seed tree method, on the other hand, may leave too few trees to interest many timber buyers.

Other Techniques

While seed tree and shelterwood systems are the most commonly employed natural regeneration systems, landowners have other options that may better fit their own particular circumstances.

- Strip clearcuts, for example, involve a clearcut timber harvest in strips 200 to 300 feet wide. Strips should be perpendicular to the direction of prevailing winds for seeds from neighboring uncut woodland to become evenly distributed. Only one cut is required in the regeneration area to allow for this kind of seed distribution.
- Seed-in-place consists of clearcutting a stand after seed fall but before the dispersed seeds can germinate. Winter logging done in a good seed year may accomplish seed-in-place regeneration.
- Finally, seeding-in-place entails clearcutting a stand 2 to 4 years following a good seed year and stand establishment. By waiting until seed germinates, a landowner can better determine the actual number of seedlings present. This method is most successful in poorly stocked stands with little or no understory competition.

The fundamentals of natural regeneration are just as important for these techniques as they are for seed tree and shelterwood systems. A bare mineral soil is necessary for a good seed catch and germination, while free-to-grow conditions are necessary for establishment and growth.

Evaluation of Natural Regeneration

Natural regeneration is successful if 300 to 700 well-distributed seedlings per acre are present and free-to-grow after three growing seasons. If all guidelines are followed, success occurs often to the point of excess. Thousands of seedlings per acre are common, a problem that can create overstocking at an early age. If stocking results in more than 900 seedlings per acre, pre-commercial thinning may be needed in years 3 to 5. Manual, chemical, or mechanical methods may be used depending on the situation. (Further discussion of precommercial thinning can be found in Chapter Ten).

If natural regeneration fails, all is not lost. Removal of seed trees and shelterwood trees followed by tree planting may be all that is required, particularly if the effects of preharvest treatments and seed bed preparation have reduced competition sufficiently. Otherwise, the landowner would need to establish a new stand through site preparation and direct seeding or planting of seedlings.

Artificial Regeneration

The planting of genetically improved seedlings onto a clearcut harvested site is at the heart of artificial regeneration and a highly successful and effective southern pine regeneration system. Although direct seeding a clearcut site is also considered artificial regeneration, only a small percentage (less than 5%) of sites are regenerated using seed. In any event, just as in the case of natural regeneration, it is extremely important that the new seedlings be planted into a site where they are free-to-grow and are not overcome by soil moisture and light competition. Therefore, typically the first step in stand regeneration is site preparation which gets the site ready for planting.

Site Preparation

The type and intensity of site preparation depends upon factors including pre-harvest stand treatments, density of woody competition left after harvest, soil type, and topography. Preparing a site preparation prescription is a job for professional foresters, with your input. Costs of site preparation may vary from around $15 per acre for burning to several hundred dollars per acre for more intensive treatments. Obviously, the objective is to choose the most cost effective technique for the job at hand. The following are several site preparation alternatives.

Fire

Fire may be used by itself in high-fuel situations or in combination with other methods of site preparation in order to clear logging slash, facilitate planting, and provide additional competition control. Although burning is simple and cheap, it results in minimal competition control and much of the site cleaning effect is temporary. As indicated previously it should be done only by trained personnel under carefully controlled conditions.

Manual

Hand tools or chain saws may be used to fell residual stems, leaving the felled stems where they fall. This method may be cost effective where scattered large diameter residuals are present. It creates little soil disturbance and can be especially useful on tracts too small for heavy equipment. Fire may be used if the fuel load is sufficient to carry a fire. Costs vary with the number and size of residual trees and may range from $50 to over $300 per acre. It is also the most dangerous method to those conducting the work.

Mechanical Methods

Mechanical methods include the following:

- Shearing—A specially designed bulldozer mounted with a "KG" blade is used to shear off residual brush and stumps close to the ground. Shearing is particularly effective against medium sized hardwoods and pine (4 to 12 inches DBH). Shearing may be followed by a burn to further clean the site and improve planting access. Many hardwood stumps, both larger and small diameters, may resprout the following summer. (Pine does not sprout.) Limit shearing to moderate slopes and stable soils to minimize erosion and soil deterioration. Cost is upwards of $100 per acre.
- Drum Chopping—A large water-filled rolling drum with external blades is pulled over brush averaging 4 inches or less in diameter. Typically a shearing blade on the front of the dozer will cut and/or push down brush which is subsequently rolled over and hopefully chopped. Again, it is common to follow this treatment with a fire to increase competition control and improve planting accessibility. Summer chopping (late June to August) is best for several reasons:
  - Resprouting is reduced
  - Going into the fall after dry weather usually results in hotter fires and better resprout kill
  - Most trees have little reserves in the roots which makes overwintering more difficult and reduces resprouting the next year

  The cost for drum chopping is around $150 per acre.
- Windrowing or Piling—In this very intensive mechanical site preparation a bulldozer with a "root rake" pushes all the logging slash, some stumps, and other heavy debris into evenly spaced windrows or piles. While this treatment leaves the site relatively free of competing vegetation, it is imperative that it be done carefully to minimize topsoil displacement. KG and regular flat blades should not be used for this operation and an experienced bulldozer operator should be used. The windrows/piles may or may not be burned. Windrowing/piling is quite expensive at around $250 per acre.
- Disking—In some cases the use of heavy disks may be all that is required to prepare a site for natural seeding or direct seeding. The site must be free of stumps and logs, however, and is most common after windrowing or piling. Cost should be around $100 per acre.

Chemical Methods

The application of herbicides is currently the most common method of site preparation for pine planting. Site preparation chemicals are low in toxicity, biodegrade relatively quickly, and are very safe when used according to the label. As with any pesticide, the user must read and follow label instructions regarding rates, timing, and application methodology. Herbicides can be broadcast by air or ground, or can be used to treat individual stems by tree injection, stump treatment, or basal spraying. Broadcast applications are the most common for site preparation and are typically followed up with a burn 4-8 weeks after application. There are a number of herbicides labeled for forestry site preparation (Table 6.3) which are highly effective at brush and weed control when used correctly.

Product	Active Ingredient	Formulation	May Control	Weak Control
Accord	glyphosate	water soluble liquid	most green plants	ashes, hickories
Arsenal	imazapyr	water soluble liquid	hardwoods, many forbs and grasses	pines, blackberry, elms
Escort	metsulfuron	dispersible granule	blackberries, manyherbaceous grassesand forbs	trees and large brush
Garlon 3A	triclopyr	water soluble liquid	hardwoods and pines	grasses, honeysuckle
Garlon 4	Triclopyr	emulsifiable concentrate	hardwoods and pines	grasses, honeysuckle
Pronone	hexazinone	granule	hardwoods, many forbs and grasses	pines, yellow-poplar
Roundup	glyphosate	water soluble liquid	most green plants	ashes, hickories
Tordon K RTU, 101	picloram	water soluble liquids	kudzu, many hardwoods, pines	grasses, hickory, ash
Velpar L	hexazinone	dispersable granuale	hardwoods, many forbs and grasses	pines, yellow-poplar
Vanquish	dicamba	water soluble liquid	forbs, woody brush	grasses
Weedone	2, 4-DP	emulsifiable concentrate	many hardwoods and forbs	grasses, kudzu

Table 6.3 Herbicides Commonly Used for Site Preparation and Timber Stand Improvement

(Mention of the products in Table 6.3 does not imply a use recommendation, nor is this list inclusive of all labeled compounds.) Chemical site preparation costs are typically around $100 per acre depending upon the chemical used, rate, and application method. The cost of the burn must be added to the application costs. Chemical prescriptions should be handled by professionals with training and experience in forestry herbicide use. For more information contact your county extension agent, state forestry agency, or consulting forester.

Site Preparation Contract

When a landowner pays for site preparation services, a written contract should be agreed upon and signed by the landowner and the provider doing the work to protect all parties involved, regardless of the site preparation method used. General items to include in a site preparation contract are:

- Maps of the site to be prepared with accompanying legal descriptions, boundary landmarks or other area descriptions
- Accurate description of the site preparation method to be used
- Method and timing of payment, cost per acre (or hour), time of year, or more specifically, beginning and completion dates
- Notification of landowner by the contractor when work begins
- Right of contractor to, or not to, subcontract to a third party
- Verification that the contractor is covered by Workmen's Compensation and liability insurance
- Provisions for settlement in case of a misunderstanding or for extension in case of inclement weather or site conditions that do not allow completion within the stated time
- Specific width, location and treatment of streamside management zones (SMZs) that should also be designated on the site maps
- Responsibility for damage to roads, fences, gates or other improvements and a clause to prevent the contractor from obstructing streams or waterways or leaving debris in roads, fields or ditches
- A clause absolving the landowner from damage to adjoining properties caused by negligence of the contractor
- Satisfactory performance guidelines, such as achieving a certain "free-to-grow" threshold or percentage
- Restriction on working during wet weather conditions when site preparation work would most likely degrade soil productivity
- Compliance with all Best Management Practices

See the Suggested readings section at the end of this chapter for more information.

Seedlings—Container Grown Versus Bareroot

There are two basic types of planting stock available for planting—bare root and container-grown seedlings. Bare root seedlings are grown in specialized forest tree nurseries and are lifted from the soil in late fall and winter and transported and planted without any soil on their roots. Container-grown seedlings, on the other hand, are similarly produced in specialized forest tree nurseries but are grown as little "plugs" 4–5 inches long, and when taken to the field, have a small amount of soil or root medium around their roots. The vast majority of southern pine planting stock are bare root seedlings with container-grown plants accounting for less than five percent of all seedlings planted annually in the South. Container grown stock is much more expensive than bare root, generally running at about $150 per thousand seedlings, while bare root is closer to $40 per thousand. Nevertheless, there may be occasion when the extra cost is justified.

- Longleaf pine is often established using container stock because bare root longleaf seedling survival is often erratic.
- Container stock can extend the planting season by allowing earlier planting in the fall or later planting in the spring.
- Because of their soil plug, container stock will usually store better than bareroot stock, needing only an occasional watering.
- Finally, container stock may do better, particularly on sites where moisture is limited.

Species Selection

A landowner should select a timber species that is well adapted to the geographic region and soil type of the planting site. Table 6.4 lists species recommended by geographical regions for many southern states. Slash pine, sand pine or other species may also be recommended. In some cases market influences should be taken into account. It is wise to seek professional forestry advice for specific recommendations for your land.

Coastal Plain	Piedmont	Mountains
Longleaf pine	Virginia pine	White pine
Loblolly pine	Longleaf pine	Virginia pine
Pond pine	Loblolly pine	Shortleaf pine
Slash pine		

Table 6.4 Generally Recommended Tree Species by Geographical Region

Planting Spacing

Tree spacing will vary with soil quality, desired rotation length, and species. Local markets will also affect spacing recommendations. In areas with poor pulpwood markets, for example, a wider initial spacing should be used so trees removed in early thinning will have greater diameters and more value such as chip-n-saw rather than pulpwood (Table 6.5).

Species	Spacing		
	Max.	Min.	Trees/Acre
Loblolly pine	10×10	6×9	435–800
Pond pine	10×10	6×9	435–800
Slash pine	10×10	6×9	435–800
Longleaf pine	7×10	6×8	622–870
Shortleaf pine	8×12	6×10	450–726
Virginia pine	8×8	6×6	675–1225
White pine	12×12	7×10	300–622

Table 6.5 Generally Recommended Initial Tree Spacings

Seedling Purchase and Care

The best season for planting the southern pines is late fall and winter, December through February. Seedlings should be purchased from specialized forest tree nurseries. Information on availability can be obtained from your local state forestry office or the Forest Landowners Association Seedling Directory. Seedlings are typically packaged in closed paper bags or in round "bales" with seedling tops sticking out both ends. Bags and bales usually contain 1000 seedlings and their roots have been sprayed with a water absorbent gel that helps keep seedling roots moist. While at the nursery, seedlings are stored in refrigerated rooms. If not, the natural heat caused by plant respiration will accumulate in the bags and reach lethal temperatures. Seedlings should be kept in refrigerated conditions if possible, and must not be allowed to heat up during transportation or storage at the planting site. Seedlings will be killed by high temperatures or drying, ruining any chance of reforestation success.

Seedling Transport

Seedling bags should be covered at all times during transportation. Transport bags in a truck with a high tarp for shade and good ventilation and a minimum of 12 inches between the tarp and the seedling bags. Freezing temperatures will kill seedling roots. If seedlings are transported during freezing weather, wrap a tarp or blanket around the bags to protect from the cold. If seedling bags must be stacked more than three deep, allow adequate air circulation between and around seedlings by using spacers such as two-by-fours to avoid overheating. Be careful not to stack seedlings so high that those on the bottom are crushed.

Field Storage

Improper seedling storage and handling at the planting site are the major cause of plantation failure. Bareroot seedlings should be planted upon receipt, especially longleaf pine seedlings which have a very short storage life (a matter of days). Nevertheless, labor availability or dry weather may not allow for immediate planting. Most winter-lifted southern pine seedlings can be stored for several weeks in cold storage (33–35° F). If cold storage is not available, seedlings should be kept in the shade in a cool, dry, well ventilated environment and used as soon as possible. Roots should be kept moist but not wet. Excess moisture can cause mold and root death. If needed, roots should be sprayed with water to moisten. Patch torn bags immediately with heavy tape to prevent seedlings from drying out. Seedlings should be inspected daily to ensure they are not drying.

Container seedlings should also be planted as soon as possible after receipt. This is particularly true if the seedlings have been "extracted" from their container and were shipped to the planting site as a small plug. Admittedly, container stock is more resistant to dessication (drying out), but still must be monitored daily to ensure the small root plugs do not dry out. Storage in a shady, protected location will help maintain adequate plug moisture. And, while container stock have some protection against freezing temperatures, the plug can freeze and kill the root system just as in the case of bareroot seedlings. If below freezing temperatures are expected (30° F or lower), then the seedlings should be covered or moved to a safer location.

Grading Seedlings

The vast majority of southern nurseries produce high quality genetically-improved planting stock that if properly cared for and given a little rain, will result in good survival and successful plantation establishment. A few nurseries grade their seedlings before shipping, but most do not. Grading is not normally needed because seedling quality is high enough that cull percentages are less than five percent and usually less than two percent. Nevertheless, a planting crew may be able to remove even these culls if given proper supervision. The following are characteristics of seedlings that may be discarded during the planting process:

- Seedlings with broken stems or with taproots broken or pruned shorter than 4 inches
- Seedlings coming from bags with a fermented smell or mold on needles
- Seedlings that have been frozen resulting in slippery bark on the root or stem
- Extremely small (less than 1/8 inch diameter at the root collar) or poorly developed seedlings (Longleaf pine should be at least 1/4 inch in diameter.)
- Seedlings whose root system have been dried by exposure to wind, sun, or high temperatures.
- Seedlings with stem galls or swelling, indicating *Cronartium fusiforme* fungal infection

Root Pruning

Bareroot seedling planting crews are tempted to prune seedling roots when they appear to be too long or excessively fibrous. Such a practice certainly makes it easier and faster to plant the seedlings, hence advantageous to the planter. Root pruning is not recommended, however, because it is very difficult to guarantee that all the root collars are lined up at the time of pruning. The result are seedlings whose taproots may be pruned too short, possibly killing the tree. Studies have consistently shown that a fibrous root system is advantageous to seedling survival and growth. Pruning them does not help the seedling, and may in fact be detrimental. If seedling roots are large, then a larger hole should be made to accommodate them. *Leaving a few fibrous roots sticking out of the hole is preferable to pruning.*

Planting

Both bareroot and container stock may be planted using hand tools or a machine planter. There are a variety of hand planting tools as well as machine planters. “Dibbles” are the most popular hand planting tool, although “hoedads” and planting shovels are also used. Planting shovels are particularly helpful for larger seedlings. Successful planting requires that a quality seedling be planted at the correct depth with roots extending straight into the planting hole. A hole 8 to 10 inches straight down is usually sufficient.

The following are several key factors to improve early growth and survival of seedlings:

- When hand planting, carry seedlings in a canvas bag or other protective holder. Do not carry bunches of seedlings by hand with the roots exposed. When machine planting, carry seedlings in a planting box on the planter. Always keep seedlings away from sunlight and drying winds.

- When removing seedlings from the shipping bag, carefully separate to prevent damaging or breaking fibrous roots. Carefully remove individual seedlings from the planting bag or box to minimize root damage.
- Insert the root system to the bottom of the hole and lift seedling to proper planting depth to give roots a downward direction.
- Plant to the correct depth:
  - Well drained sites (sand and sandy loams)—plant root collar 2 to 3 inches below ground line except for longleaf pine which should be planted at the root collar
  - Poorly drained sites (clay and silt soils)—plant root collars 1 inch below ground line
  - Containerized seedlings may be planted to the same depth as bareroot seedlings. If the planting tool does not allow this, then plant the plug at, or slightly below, the top of the plug.
- Close hole properly to ensure root contact with the soil. Check by pulling the tops of seedlings to determine that seedlings are held "snugly" in the planting hole. Make sure the hole is firmly closed and there is not an air cavity at the bottom of the hole. Avoid areas of loose soil or organic matter.
- Maintain quality control by frequently checking seedling condition, planting depth, and proper packing of soil.

Tree Planting Contracts

A written tree planting contract protects both the landowner and contractor from misunderstandings and possible legal proceedings. Include specific provisions covering the following items in a tree planting agreement:

- Maps of the site to be prepared with accompanying legal descriptions, boundary landmarks or other area descriptions
- Acres, price per acre, species, seedling source, spacing, and planting method
- Time of planting, preferably a starting and ending date
- Designation of responsibility for obtaining, transporting, storing, and handling seedlings (to landowner or contractor)
- Satisfactory performance guidelines, such as achieving a certain minimum survival percentage and designation of a person(s) responsible for and date of seedling survival count (generally the contractor should be liable to replant—at no cost to you—areas which fall below an agreed percentage survival specified in the contract)
- Method and timing of payment, cost per acre (or hour), time of year, or more specifically, beginning and completion dates
- Verification that the contractor is covered by Workmen's Compensation and liability insurance
- Right of contractor to, or not to, subcontract to a third party
- Specify, if necessary, the use of insecticide treated seedlings for *Pales* weevil control
- Site access provision for planting crew

- Provision requiring the contractor to contact the landowner upon commencement of planting
- Provision for settlement of disputes

Contract rates for machine planting cutover land average about $90 per acre and about $50 per acre for hand planting.

Direct Seeding

An alternative to planting seedlings is spot or broadcast distribution of seed. Newly cutover sites can be regenerated for one half to one third the cost of planting, provided the scattered seed falls on mineral soil and a competition-free site so it can germinate and grow. Commercial seed companies and some timber companies sell seed ready to sow. Follow these guidelines for successful direct seeding:

- Use seed treated with a bird, mammal and insect repellent.
- Use stratified seed for spring planting (February, March). Stratified seed has been stored for 45 to 60 days in moist conditions at 36° to 40° F which is necessary to prepare seed for germination.
- Unstratified seed is better for early season (November, January) seeding.
- Broadcast 1/2 to 3/4 pound of seed per acre by air or by hand cranked seeder.
- Spot seed 1/4 to a pound of seed per acre at points determined by the desired spacing. At the same time, choose spots where chances of success are greatest. Three or four seeds should be used per spot.

Similar to natural regeneration, direct seeded stands are more susceptible to drought and early competition pressure than are planted stands. A systematic inventory of seedling survival should be done at the end of the first and second growing seasons. Seed costs range from $15 to $25 per pound.

Summary

Artificial and natural methods of reforestation can be successfully used to reforest pines in southern timberlands. Each method has advantages under certain situations. Natural regeneration is a relatively inexpensive alternative, but has many risks associated with it. Artificial regeneration, while more expensive, is more likely to maximize volume production in the shortest time. There are typically many facets to consider and alternatives to compare. Landowners will typically benefit from consulting state forestry agencies, forestry consultants, industrial foresters, or other professional foresters when contemplating their pine regeneration alternatives.

Suggested Resources

Readings

- Dangerfield, C. W. and D. J. Moorhead. 1997. "Evaluating pine regeneration economic opportunities." *Forest Landowner* V57 No.5. September/October. pp 12–15.
- Edwards, M. B. 1987. "Natural regeneration of loblolly pine." USDA Forest Service. Southeastern Forest Experiment Station. *General Technical Report SE-47.*
- Hamilton, R. A. 1996. "Site preparation methods and contracts." *Publication WON-15.* North Carolina Cooperative Extension Service, North Carolina State University. Raleigh, NC.
- Hamilton, R. A. 1997. "Steps to successful pine plantings." *Publication WON-16.* North Carolina Cooperative Extension Service, North Carolina State University. Raleigh, NC.
- Moorhead, David J. 1989. "A Guide to the Care and Planting of Southern Pine." *Management Bulletin R8-MB39.* USDA Forest Service, Southern Region. Atlanta, GA.
- Wade, D. D. and J. D. Lunsford. 1989. "A Guide for Prescribed Fire in Southern Forests." *Technical Publication R8-TP 11.* USDA Forest Service, Southern Region. Atlanta, GA.
- Williston, H. L., and W. E. Balmer. 1983. "Direct Seeding of Southern Pines: A Regeneration Alternative." USDA Forest Service, Southern Region. Atlanta, GA.

Chapter 7: Regeneration of Southern Hardwoods

When considering hardwood regeneration and management, you should pay attention to stand conditions, soil-site types, species and various regeneration methods. This chapter examines how to create or encourage new hardwood stands. As will become apparent, regenerating hardwoods can involve actions well before a "final" harvest. Obtaining desirable regeneration is a *process,* not merely an event that follows the harvest. Thus, the topic of regenerating hardwoods overlaps the topic of intermediate hardwood stand management, which is covered in Chapter 11.

Stand Conditions

The first decision to make is whether to manage the existing hardwood stand or regenerate a new one. Most southern hardwood stands are dominated by low-quality trees resulting from a history of repeated partial harvests. Previous harvests have likely removed high-quality trees while leaving increasingly poor quality trees, a practice called "high grading."

A minimum of 60 well-spaced, good quality trees of desirable species per acre is needed to make management of the existing stand practical on most sites. This assumes an approximate spacing of 27 feet between crop trees that are straight, relatively knot-free, and free of insect or disease damage. Desirable species will vary across the South, but generally the oaks, ashes, yellow-poplar, sycamore, pecan, sweetgum, tupelo-gum and walnut are examples of quality species. Managing existing stands will be discussed in Chapter 11.

If a manageable stand is not present or if the current stand is mature, a regeneration plan should be implemented. Figure 7.1 shows a generalized prescription for assessing the regeneration of desirable species in hardwood stands. No cookbook treatment for hardwood regeneration can totally account for the numerous biotic and abiotic factors which affect regeneration potential (e.g., site species mixtures, past disturbances, flooding, fire, grazing, etc.).

Natural Regeneration Methods

Hardwood reproduction will come from coppice (stump and root sprouts), seed or advanced seedlings (Table 7.1). In stands with tree stump diameters smaller than 12 inches, coppice will be the main source of reproduction, whereas seed or seedlings will dominate in larger diameter stands. The original stand should be harvested from November through March, if possible, as this favors coppice development and also takes advantage of current-year seed crops.

Regeneration methods commonly used with hardwoods are single-tree selection, group selection, shelterwood, seed tree and clearcut. The merits of each method are discussed, together with recommendations for their applicability to the southern hardwood forest.

Single-Tree Selection

Single-tree selection has a sound theoretical basis but is impractical to apply except in special situations where landowners can apply the muscle and continual attention needed to make the system work. Single-tree selection is best adapted to regeneration of shade tolerant species and in all-age stands. The goal is to maintain a similar basal area in each age class; thus many trees exist in small-diameter classes and few trees exist in large-diameter classes.

At about 10-year intervals, the largest trees are harvested. Then the poorer-quality trees and less desirable species are removed down through the diameter size classes to the smallest trees. Each harvest strives to create conditions favorable for desired reproduction, to enhance growth through crop tree release, and to maintain desired stand structure.

Variations on this method have been advocated and employed in regenerating southern hardwoods. Generally such efforts have met with poor success because the highest value and largest trees are harvested without control of small and undesirable trees, leaving a progressively poorer stand with each stand entry. To be successful, the forester must remove trees of poor quality as well as some precommercial trees to create openings favorable for regenerating shade intolerant species. The method has value, when properly applied in situations that preclude complete overstory removal, such as stream borders, recreation areas and locations where visual quality is a prime consideration. However, getting operators to exercise the care required in such a system can be difficult, and the costs involved can be high.

Pre-harvest Regeneration Evaluation for Desirable Species

1. Stump Sprout Potential
2. Advance Reproduction
3. Windblown Seed from Adjacent Trees
4. Seed Accumulating in Forest Duff

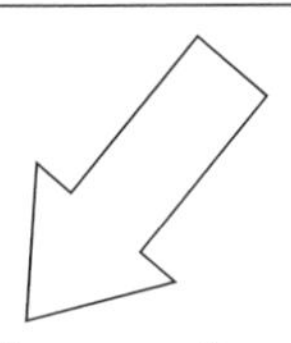

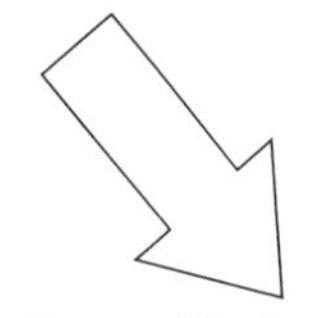

Desirable Regeneration Prospects <u>POOR</u> (<200 potential stems/acre)

Options

1. Take steps to promote advance reproduction by increasing light to the forest floor through understory removals and/or partial overstory cuttings
2. Supplemental planting
3. Artificial regeneration (plantings)

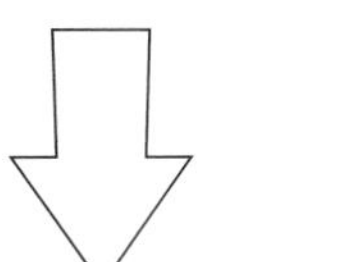

<u>Less</u> than 200 free-to-grow seedlings of desirable species

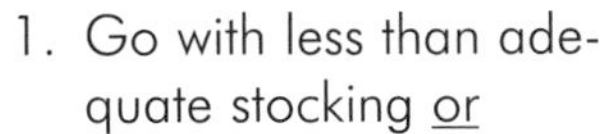

1. Go with less than adequate stocking <u>or</u>
2. Convert to plantation

Desirable Regeneration Prospects <u>GOOD</u> (>200 potential stems/acre)

Procedures

1. Treat and harvest during dormant season
2. Control residual stems prior to next growing season

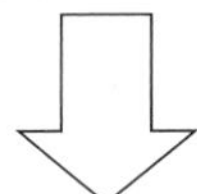

<u>Evaluate at Age 3</u>

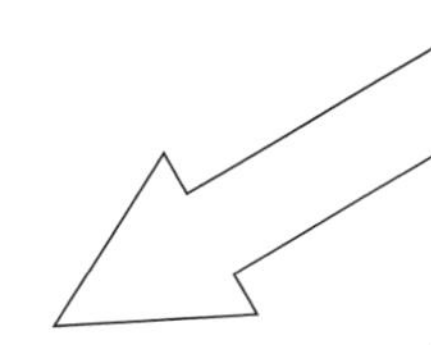

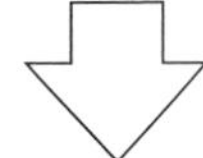

<u>More</u> than 200 free-to-grow seedlings of desirable species

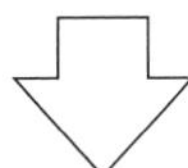

1. Leave alone <u>or</u>
2. Clean, weed or thin as needed

Figure 7.1. Procedures for assessing regeneration of desirable species in hardwood stands.

Relative Importance of Reproduction Source in Relation to Species Regeneration Potential[a]						
Species	Seed from Current Seed Crop	Seed Stored in Forest Floor	Advance Reproduc.[b]	Stump Sprouts[c]	Root Sprouts (Suckers) from Cut Trees	Shade Tolerance
Hardwoods						
American basswood			1	1	2	tolerant
American beech			1	3	2	very tolerant
American elm			1	3	2	intermediate
American holly		1	2	2		very tolerant
American hornbeam			1	2		very tolerant
Black cherry		1	2	2		intolerant
Blackgum			1	2		tolerant
Black locust		2			1	very intolerant
Black walnut	1		1	2		intolerant
Black willow	1			2	2	very intolerant
Boxelder	2		1	1		tolerant
Buckeye	1	1	2	2		tolerant
Cucumbertree		1	2	2		intermediate
Eastern cottonwood	1			1	3	very intolerant
Eastern hophornbeam			1	2		very tolerant
Eastern redbud		2	1	1		tolerant
Flowering dogwood			1	2		very tolerant
Green ash		1	1	2		intermediate
Hackberry			1	3		intermediate
Hickories[d]			1	3		intermediate
Oaks[e]			1	2		intermediate
Persimmon		2	1	1	2	very tolerant
Red maple			1	2		tolerant
River birch	1		2	2		intolerant
Sassafras			1	2	1	intolerant
Silver maple			1	2		tolerant
Slippery elm			1	3	2	tolerant
Sourwood		2	1	1		tolerant
Sugar maple			1	2		very tolerant
Sweetgum	1			1	1	intolerant
Sycamore	1		2	2		intolerant
White ash		1	1	2		intermediate
Yellow birch	1		2			intermediate
Yellow-poplar		1	2	2		intolerant

[a]1=primary source; 2=potentially significant but not primary source; 3=minor source. Relative importance of reproduction source is for sawtimber-size stands.
[b]Includes seedlings, seedling sprouts and root sprouts.
[c]Sprouts originating from stumps of trees >= 2 inches DBH.
[d]Hickories as a genus are mostly intermediate in shade tolerance. Mockernut and bitternut hickories range toward intolerant.
[e]Oaks as a genus are mostly intermediate in shade tolerance. Red oaks range more toward the intolerant scale.

Source: Johnson, 1989; Burns and Honkala, 1990; Putnam et al, 1960

Table 7.1. Sources of Reproduction and Shade Tolerance of Some Species in Hardwood Forests

Group Selection

Group selection has the same theoretical basis for implementation as single-tree selection. The major difference is that trees are removed in groups from appropriate diameter classes, which creates larger openings than does removing single trees. Openings created typically range from 1/10 to 1/2 acre, with larger openings desired to decrease shade effects.

Patch clearcutting 1 to 3 acres is an extension of group selection. Such cuttings foster reproduction comparable to that of clearcuts and provide considerable stand diversity. The method has greater application than single-tree selection for environmentally sensitive areas.

Shelterwood

Shelterwood is removal of approximately 50 percent of the stand in the first of two harvests. It is reputed to be the most economical method for regenerating hardwoods of medium shade tolerance. Experience shows that ensuing reproduction is comparable to that from clearcuts. However, development of shelterwood reproduction is hindered by the overstory trees, causing a shift to more shade tolerant species composition not encountered in clearcuts. A solution to the problem is removal of the overstory stand as soon as the desired reproduction is obtained. However, this removal may pose several harvest problems (see Chapters 11 and 12).

Procedures for implementing the shelterwood method include a preliminary cut from 3 to 10 years before final harvest to establish advanced reproduction from seed, seedling, or sprouts. The preliminary cut should be from "below," removing the smallest trees first, and then advancing to the larger trees. Undesirable trees may be controlled by poison injection or girdling if the cost of harvesting is prohibitive. The objective is to leave an overstory canopy of desirable trees to cover about 50 percent of the area.

Using the shelterwood method in southern hardwoods affords latitude in creating desired regeneration conditions under partial shade. The latitude is regulated by the severity of cutting and by the degree of disturbance of the forest floor. The method is relatively easy to implement, and tall-tree cover can be maintained for many years to minimize visual objections. In visually sensitive areas, final removal can be spread over two cuts, which can nearly eliminate the visual presence of tops, slash, and stumps. Care must be taken during intermediate cuts to avoid soil puddling and compaction and to prevent damage to the residual trees.

A modification of the shelterwood method, known as "shelterwood with reserves" or "deferment cutting" results in a two-aged stand. The method involves the removal of all but 15 to 20 trees per acre, averaging 14 inches in diameter. These remaining trees are termed reserves or standards. The limited number of reserve trees allows abundant light to reach the forest floor, provides for the rapid growth of the understory, and the development of two age classes. The treated area is subjected to site preparation treatments similar to what might be done in clearcut situations (see clearcut discussion below) to promote desirable species. In contrast to the shelterwood method, the reserve trees are left standing for a second rotation, thus maintaining two predominant age classes. The limited number of reserve trees ensures continued development of the regenerating age class.

Two-aged methods have the ability to deliver several important advantages compared to other regeneration cuts commonly employed (such as clearcutting):

- Maintenance of sexual reproduction throughout the rotation
- Maintenance of advance reproduction development throughout the rotation
- Reduced visual or aesthetic impact

However, two-aged methods also pose substantial risks.

- Some species have a tendency to produce epicormic branches, which decreases the future value of the tree.
- Solitary trees are much more susceptible to storm damage.
- Reserves should be of an age where they will prosper for a second rotation (another 70-90 years). Often these reserves are already mature or excessively mature and they will not survive through an additional rotation.

The choice of prescribing two-age methods will depend on the tradeoffs between timber production, obtaining and developing desirable regeneration, aesthetics and risks associated with maintaining large trees for an additional rotation length.

Seed Tree Cuts

Seed tree cuts have little practical value in regenerating southern hardwoods. The purpose of seed trees is to provide seed for a new stand. However, these "new" seeds are rarely needed to regenerate hardwood stands.

Clearcut

Clearcut has the widest applicability of any method for natural regeneration of southern hardwoods. The clearcut method of fostering natural reproduction mimics nature better than any regeneration method. Silvicultural clearcuts, where growth of residuals larger than 1.5 inches DBH is inhibited, as opposed to economic clearcuts, where no residual control is practiced, have proven valuable on every forest site type. The method favors the shade-intolerant species (sun-loving) which are desired for timber production. It is especially applicable for refurbishing high graded stands of high potential productivity. Since the majority of southern hardwood stands are in a degraded condition, clearcutting is recommended over other regeneration methods except in environmentally sensitive areas.

Chances for achieving desired natural hardwood regeneration from clearcutting can be improved by adhering to the following guidelines:

- Schedule harvests, if possible, from November through March to take advantage of current-year seed crops and to favor coppice regeneration.
    - Coppice reproduction emanating from April through October harvests is reduced in quantity and quality compared to coppice from dormant season harvest.
    - Soil degradation from logging and residual control must be minimized at all times and especially during the dormant season when soils are typically water-saturated.
- Harvest the parent stand as completely as possible.
- Favor logging systems that damage or destroy the above-ground portion of all residuals, including advanced reproduction. This allows shade intolerant species to receive adequate sunlight to enhance survival and early growth.
- Favor logging systems that maximize soil scarification without causing soil degradation.
- Harvest trees at a stump height less than 12 inches when practical.
- Prevent water impoundment on bottomland sites by constructing roads and keeping water channels free of slash and debris.
- Minimize erosion on upland sites by constructing proper roads and by orienting skid trails with terrain.
- Control residuals larger than 1.5 inches DBH by shearing, chopping, felling, girdling, or with chemical herbicides within six months of harvest when economically feasible, to obtain benefit from prevailing seed crop and from soil scarification resulting from logging.
- Remember that shearing is the preferred method for obtaining natural regeneration. Piling, following shearing, is not recommended except where vines are prolific. It will reduce quantity and quality of desired reproduction compared to shearing only.
- Control residuals by girdling and with herbicides and by shearing and piling where heavy vine cover occurs.
- Consider that chopping fosters poor quality coppice regeneration from root- sprouting species such as sweetgum and black locust. It also causes rampant spread of Japanese honeysuckle and other undesired vines.

Hardwood Plantations

While hardwood plantations are not as common as pine plantations, hardwood plantations may be used to reclaim disturbed landscapes or abandoned farm fields or when a specific species is desired. As a first step, a soil survey should be employed to determine soil type, texture and drainage for each site, followed by determining which species or combination of hardwood species is best suited to the particular site. Once the site has been evaluated and the species of hardwood is matched to the site, clear objectives for your hardwood plantation must be established. For example, you should decide whether you wish to produce hardwood sawtimber or wildlife habitat.

Grading Seedlings

The better the quality of the hardwood seedlings planted, the greater the chance for plantation success. The following are some characteristics to look for when selecting hardwood seedlings:

- Tap roots should be 6–10 inches long.
- Tops should be 18 inches tall; 24 inches is better.
- Root collar diameter should be at least 3/8-inch for all species.
- Insist on a minimum of 5–7 first order lateral roots for oak and 7–9 for walnut, larger than 1 mm in size at the base.
- They should have been grown at low bed densities (3–6 seedlings per square foot).

Seedling Storage

Seedlings should be planted directly upon receipt, however, changing weather conditions and planting schedules can delay planting. The storage of seedlings in a barn or shed is acceptable provided the seedlings remain shaded, and temperatures do not drop below freezing or above 50° F. Should planting be delayed for more than two weeks, place seedlings in cold storage (33°–40° F) or delay pick-up from the nursery.

Seedling Handling

Improper handling of seedlings is a major cause of plantation failure. The following are guidelines in handling seedlings:

- Soak seedlings 2–6 hours before they are planted for maximum seedling moisture content.
- Remove only those seedlings that will be planted that day.
- Trim first-order lateral roots to 3- to 4-inch lengths prior to planting.
- Carry seedlings in planting bags or buckets to keep the roots moist.
- Do not carry seedlings by hand or expose roots to wind and sun.

Seedling Planting

Site preparation can be accomplished by mowing, disking or burning (see Chapter 5). Generally, the probability of successfully establishing hardwoods is higher with more intense site preparation. Planting should occur during the dormant season (January–March). Hardwood seedlings can be planted by hand (using a KBC dibble, planting spades, post-hole diggers and power augers) or machine. If planting by machine, ensure that the seedling is properly placed in the hole. The following factors will increase the success of hardwood plantings:

- Planting holes should be 6–8 inches in diameter and 8–12 inches deep.
- Place seedlings in hole so the root collar is slightly below the soil surface.
- Do not bend, ball or curl the root tip in the bottom of the hole.
- Close the hole completely by tightly packing the soil around the roots.

- Spacings should range between 10 × 10 feet to 20 × 20 feet. One-hundred to 450 trees are usually planted per acre.

Weed Control

Competition should be controlled around seedlings for the first 2 to 5 years or until crown closure. Weeds can be controlled after planting by mowing, disking or herbicide application. Contact a consulting forester, county extension agent, or state forester for advice about herbicides. Strongly consider contracting herbicide treatment to a competent professional. If the landowner applies the herbicide, ensure all restrictions, laws, and manufacturer's instructions are carefully followed.

Soil-Site Types

Descriptions of the different soil-site types where hardwoods occur are outlined below.

Muck Swamp

Very poorly drained area, usually with standing water. Broad expanses between tidewater and upstream runs and along blackwater rivers and branch bottom stands. Characterized by heavy accumulation of organic matter (amorphous, lacking structure). Soils range from silt loam through clay. Water tupelo and bald cypress are common in deeply flooded areas and swamp black gum predominates toward the fringes.

Peat Swamp (Headwater Swamp)

Broad interstream areas from which blackwater rivers and branch bottoms originate; poorly drained, with heavy accumulations of raw organic matter. Soils resemble those of muck swamps but in general are heavier and of better site quality. Examples are Dismal Swamp of Virginia-North Carolina; Green Swamp of North Carolina-South Carolina; Little Wambaw of South Carolina; and Okefenokee Swamp of Georgia-Florida. Swamp black gum and red maple predominate in mixture of many hardwood species; coniferous species found are spruce, slash, loblolly and pond pines, and Atlantic white cedar to the north.

Wet Flat

Topographically similar to peat swamps, but better drained because of higher elevation. The nonalluvial soils may possess some accumulation of organic matter but fertility is superior to peat swamps because of superior parent material. Abandoned rice fields of the southern lower coastal plain often fall in this category. Species include:

- On wetter portion—sweetgum; red maple; water, laurel and willow oaks; ashes; loblolly and slash pine; elms and other species
- On "islands" with better drainage—cherrybark, Shumard and swamp chestnut oaks, yellow-poplar, hickories, and occasionally beech

Slightly elevated areas of wet flats are known as "pocosins," in the North Coastal Plain, whereas in the South Coastal Plain they are called "hammocks" or "hummocks." Coniferous species and bays dominate poccosins, while hammocks are composed of mixed hardwoods.

Red River Bottom

This is the floodplain of major drainage systems originating in the Piedmont or mountains. Some organic matter may accumulate on the clay soils. Water tupelo dominates over cypress, red maple, swamp blackgum, swamp cottonwood, laurel oak and others. On first bottoms (low ridges) which flood periodically to considerable depths, drainage is fairly rapid because of high elevation. Soils are loam or silt loam. Species include sweetgum, green ash; water hickory; sycamore; red maple; river birch; elms; and willow, water, laurel, and overcup oaks. At higher elevations, second bottoms and terraces (high ridges and fronts) are found. Flooding is infrequent or rare, and species of cherrybark, swamp chestnut, white oaks, hickories, beech, and occasionally yellow-poplar make their appearance.

Black River Bottom

These are floodplains of major water systems originating in the Coastal Plain. Classification of minor site types and species similar to red river bottom, with exception of muck swamps being more prevalent and first and second bottoms and terraces being on a more modest scale.

Branch Bottom

Relatively flat, alluvial land along minor drainage systems which are subject to minor overflow. On wetter portions with heavier soils, the predominant species are willow, water, and laurel oaks, swamp black gum, sweetgum, red maple, and ash. The lighter soils of second bottoms and terraces support cherrybark, shumard, swamp chestnut, and white oaks, sweetgum, hickory, yellow-poplar, slash, spruce, and loblolly pines. Sloughs and oxbows of limited extent along the main channel support tupelo and swamp blackgum.

Bottomland

In lower Piedmont, conditions identical to red river bottom are encountered. However, upstream, sloughs, oxbows and first bottoms decrease in frequency and area until only well-drained bottomland (second bottom and terrace) is encountered. This type also includes the area adjacent to the tertiary streams of the red rivers. Species found here are sycamore, yellow-poplar, sweetgum, green ash, cottonwood, water and willow oak, loblolly pine, and others.

Major and Minor Bottoms

Major and minor bottoms may be distinguished by the deposition and mineralogy of sediments. The alluvium deposited in major river bottoms, such as the Mississippi River, may come from hundreds and even thousands of miles away. Major bottoms are composed of materials from all textural classes (sand, silt, and clay) and several kinds of minerals. In contrast, the alluvial deposits of minor bottoms are of local origin and vary much less in textural class and mineralogy. Although the soil/site topographic features are similar between the Atlantic and Gulf Coastal Plains, the deposition and the accompanying species composition can be quite different. A summary of species suitability by site classification for the minor and major bottoms is listed in Table 7.2.

Coves, Gulf and Lower Slopes Adjacent to Streams

This site type is characterized by uneroded, fertile mineral soils that are moist but well drained. It is common to upper Piedmont and mountain provinces. In the Cumberland Plateau range, valleys, or hollows extending from the crest of the range to major drainage systems are known as gulfs. They are characterized by plateaus that are separated by steep terrain. Here the mixed mesophytic (medium conditions of moisture) forest consisting of yellow-poplar, sweetgum, white ash, northern red oak, black oak, black cherry and black walnut is at its best.

Upland Slopes and Ridges

Here the mineral soils are shallower and dryer than the coves and lower slopes. White oak and hickory predominate but yellow-poplar, black and scarlet oaks, and blackgum are also common. Shortleaf and Virginia pines are common coniferous associates.

Summary

Successful natural regeneration of southern hardwoods can be accomplished by assessing stand conditions and adhering to accepted practices. Viable coppice reproduction cannot usually be expected from trees larger than about a 12-inch stump diameter.

Single tree and group selection regeneration methods can be applied in southern hardwood stands where timber production in conjunction with maintaining high visual quality is a goal and where costs can be justified or ignored. The shelterwood regeneration method fosters reproduction comparable to that of a clearcut. Overstory trees should be removed as soon as the reproduction becomes established to prevent reversion to shade-tolerant species.

The regeneration method most widely advocated for southern hardwoods is clearcutting. Good results can be obtained by removing all merchantable timber and controlling residuals. Shearing residual trees to a stump height less than 12 inches within six months after harvest and, if possible, from October through March gives best results. Girdling, applying chemical herbicide, chopping, and burning are viable alternatives for controlling residuals in areas with high water tables or precipitous slopes.

Topographic	Desirable/Suitable Species	
Site Position	**Major Bottoms**	**Minor Bottoms**
Bars	Cottonwood, willow	River birch, willow
Fronts	Cottonwood, water oak, sweetgum, sycamore, pecan, green ash (swamp chestnut oak, cherrybark oak)[1]	Cherrybark oak, Shumard oak, sweetgum, sycamore, yellow-poplar
Ridges	Water oak, willow oak, sweetgum, green ash (cherrybark oak, swamp chestnut oak)[1]	Cherrybark oak, Shumard oak, swamp chestnut oak
High Flats	Nuttall oak, green ash, persimmon, sugarberry	Cherrybark oak Shumard oak, water oak, willow oak, swamp chestnut oak
Low Flats	Overcup oak, water hickory, green ash, persimmon, sugarberry	Willow oak, overcup oak, green ash, persimmon
Sloughs	Overcup oak, water hickory, black willow	Overcup oak, persimmon
Swamps	Baldcypress, water tupelo	Baldcypress, water tupelo, swamp tupelo

[1]Species in parenthesis may not be found on those sites in the Mississippi River floodplain

Adapted from: Ezell and Hodges, 1995

Table 7.2. Site Suitability for Bottomland Hardwoods in Major and Minor Bottoms

Suggested Resources

Readings

- Burns, R.M. and B.H. Honkala. 1990. "Silvics of North America. Agriculture" *Handbook 654.* USDA Forest Service, Washington, DC. (2 volumes). *www.srs.fs.usda.gov/pubs/misc/ag_654_vol2.pdf*
- Clatterbuck, W.K. and J.S. Meadows. 1993. "Regenerating oaks in the bottomlands." In: Loftis, D.L. and C.E. McGee (eds.), *Oak Regeneration: Serious Problems Practical Recommendations: Symposium Proceedings,* 8-10 September 1992, Knoxville, TN. General Technical Report SE-84. USDA Forest Service, Southeastern Forest Experiment Station, Asheville, NC. *www.srs.fs.usda.gov/pubs/gtr/gtr_se084.pdf*
- Ezell, A.W. and J.D. Hodges. 1995. "Bottomland hardwood management: Species/site relationships." *Publication 2004.* Mississippi State Cooperative Extension Service, Mississippi State University. Starkville, MS. *http://msucares.com/pubs/publications/p2004.htm*

- Hodges, J.D. 1989. "Regeneration of bottomland oaks." *Forest Farmer* 49(1):10–11.
- Johnson, P.S. 1989. "Principles of natural regeneration." *Central Hardwood Notes Publication 3.01.* USDA Forest Service, North Central Experiment Station, St. Paul, MN. *www.ncrs.fs.fed.us/pubs/ch/ch_3_01.pdf*
- Meadows, J.S. and J.A. Stanturf. 1997. "Silvicultural systems for southern bottomland hardwood forests." *Forest Ecology and Management* 90:127–140. *www.srs.fs.usda.gov/pubs/ja/ja_meadows008.pdf*
- Perkey, A.W., B.L. Wilkins, and H.C. Smith. 1994. "Crop tree management in eastern hardwoods." NA-TP-19-93. USDA Forest Service, Northeastern Area State and Private Forestry, Forest Resources Management, Morgantown, WV.
- Putnam, J.A., G.M. Furnival, and J.S. McKnight. 1960. "Management and inventory of southern hardwoods." *Agriculture Handbook 181.* USDA Forest Service, Washington, DC.
- Robison, D.J., R.E. Bardon, F.W. Cubbage, D. Frederick, C. Moorman, J.L. Schuler, C.A. Harper, and J. Siry. 2004. "Management approaches for hardwoods in the South." *Forest Landowner* 63(2):5.
- Sims, D.H. and D.L. Loftis. 1989. "Regenerating northern red oak on high quality sites." *Forest Farmer* 49(1):12–13.
- USDA Forest Service, Southern Region. 1994. "Southern Hardwood Management." *Management Bulletin R8-MB 67.* Atlanta, GA. *www.sref.info/publications/online_pubs/hardwood_management.pdf*

Review Questions

1. How do hardwood trees regenerate themselves?

2. What is high grading?

3. What are the methods of regeneration? Which method is recommended the most?

4. Distinguish between major and minor bottoms. Distinguish between red river bottoms and black river bottoms.

5. Why should seedlings be graded?

6. Many seedlings are stressed during storage, handling and planting operations. What procedures are recommended to ensure survival of seedlings during these operations?

7. Why is it advantageous to harvest your timber during the dormant season—November through March?

Chapter 8: Finance, Taxes and Investment Issues

Many forest owners have management objectives that include financial gain as well as allowing the landowner the opportunity to manage their forest for wildlife habitat, esthetics, recreation and other uses. Historically, income from managed timber stands has fared well with agricultural crops in terms of return per acre per year. Even managed pre-merchantable timber stands have proven to increase the property value of forestland substantially in comparison to bare or unmanaged cutover woodland. Financial returns vary widely due to the inherent productivity of the land (measured by site index), stand conditions, tree species, the availability and vigor of timber markets and intensity of management practice (use of herbicides, fertilizer, etc. to increase production). Availability of financial incentives and other factors also affect the return on investment for forest owners. This chapter provides an overview of current financial, tax and investment issues and opportunities that affect forest owners today.

The information which follows applies to federal cost-share and income tax laws as of the writing of this text. Consult your county forester, Extension agent, private consulting forester and/or tax specialist for recent updates or referrals. For a complete discussion of federal timber taxes contact the Forest Landowners Association or refer to the suggested readings at the end of this chapter. Information on cost-share programs may be obtained from the United States Department of

Agriculture (USDA) Natural Resources Conservation Service (NRCS), the USDA Farm Service Agency (FSA), and the state forestry agency local offices.

Cost-Share Programs for Establishment and Management

Numerous state and federal programs are in place to encourage active forest management due to the proven economic and environmental benefits to society. Cost-share programs can partially defray costs of initial planting or replanting after harvest. Some cost-share programs cover part of the costs for forest improvement practices such as hazard reduction burning, wildlife habitat improvement or streamside management as well as other intermediate stand practices.

Many Southern states have state cost-share programs for forest practices. For information about programs in your state, visit *http://www.srs.fs.usda.gov/econ/data/forestincentives/*. In most states the state programs are administered by state forestry agencies or associations. Under these programs, a landowner is partially reimbursed for the costs of site preparation, seedlings, tree planting, release of desirable seedlings from competing vegetation, or other work needed to establish a new forest. A forest management plan approved by the state forestry organization or agency is often required to qualify for this assistance.

Every southern state has opportunities for cost-share programs administered by federal agencies such as the NRCS or the FSA. One of the major federal programs, the Conservation Reserve Program (CRP), established by the 1985 Food Security Act (Farm Bill), is expected to ultimately retire tens of millions of acres of highly erodible, marginal cropland nationwide. Landowners may use the retired cropland to grow trees as permanent wildlife habitat, for permanently growing introduced or native grasses and legumes or combinations of permanent covers. Sign-up in the CRP is limited to active or retiring farmers who successfully bid into the program.

An Emergency Forestry Conservation Reserve Program (EFCRP) was created in response to the forest destruction caused by Hurricanes Katrina, Rita, Dennis, Ophelia, and Wilma. This program is limited to landowners with less than 500 acres of private non-industrial forestland, with damage to at least 35% of merchantable timber stands. Details may be obtained from the local Farm Service Agency in the affected areas.

For these programs the FSA will reimburse up to 50 percent of the cost of establishing permanent covers and will pay an additional rental fee over a 10-year period to participating landowners. Retired acreage may not be grazed, harvested or used in any commercial manner other than for hunting leases during the 10-year period. There are also repayment penalties if the land is sold or the timber is removed before the contract has expired. In certain cases, the CRP may be extended for additional time periods. Landowners may sign up for the program during open enrollment periods at the county FSA office. Visit *http://www.fsa.usda.gov/FSA* for more information on enrollment period announcement.

At the time this publication went to press, the Forest Land Enhancement Program (FLEP), a program that resulted from the 2002 Farm Bill, had not been fully funded. This program was designed to combine numerous cost-share programs of the past along with numerous new incentives for landowners to enhance and improve their forests for a variety of benefits. Two earlier cost-share programs, the Forestry Incentives Program (FIP) and the Stewardship Incentives Program (SIP) were sunset when this new program was authorized. The President's proposed 2008 Farm Bill includes a consolidation of cost-share programs. It is likely that whatever Farm Bill is passed will have significant changes in cost-share programs, including the conservation programs.

Other federal conservation-oriented cost-share programs worthy of investigation for their potential tree planting/forest management cost-share assistance include the Environmental Quality Incentives Program (EQIP), the Wildlife Habitat Incentives Program (WHIP) and the Wetlands Reserve Program (WRP). The Private Landowner Network has an excellent listing of various cost-share and grant programs for landowners with links to appropriate agencies. Their web address is *http://www.privatelandownernetwork.org/*.

EQIP was reauthorized in the 2002 Farm Bill. EQIP offers contracts with a minimum term that ends one year after the implementation of the last scheduled practices and a maximum term of ten years. These contracts provide incentive payments and cost-sharing to implement conservation practices. Persons who are engaged in livestock or agricultural production on eligible land may participate in the EQIP program. EQIP activities are carried out according to an environmental quality incentives program plan of operations developed in conjunction with the producer who identifies the appropriate conservation practice or practices to address the resource concerns. The practices are subject to USDA NRCS technical standards adapted for local conditions. The local conservation district approves the plan.

EQIP may cost-share up to 75 percent of the costs of certain conservation practices. Incentive payments may be provided for up to three years to encourage producers to carry out management practices they may not otherwise use without the incentive. However, limited resource producers and beginning farmers and ranchers may be eligible for cost-shares up to 90 percent. Farmers and ranchers may elect to use a certified third-party provider for technical assistance. An individual or entity may not receive, directly or indirectly, cost-share or incentive payments that, in the aggregate, exceed $450,000 for all EQIP contracts engaged during the term of the Farm Bill.

The Wildlife Habitat Incentives Program (WHIP) is a voluntary program for people who want to develop and improve wildlife habitat primarily on private land. Through WHIP, USDA's Natural Resources Conservation Service provides both technical assistance and up to 75 percent cost-share assistance to establish and improve fish and wildlife habitat. WHIP agreements between NRCS and the participant generally last from 5 to 10 years from the date the agreement is signed.

By targeting wildlife habitat projects on all lands and aquatic areas, WHIP provides assistance to conservation minded landowners who are unable to meet the specific eligibility requirements of other USDA conservation programs. The Farm Security and Rural Investment Act of 2002 reauthorized WHIP as a voluntary approach to improving wildlife habitat in our nation. Program administration of WHIP is provided under the NRCS.

The Wetlands Reserve Program (WRP) is a voluntary program offering landowners the opportunity to protect, restore, and enhance wetlands on their property. The USDA NRCS provides technical and financial support. The NRCS goal is to achieve the greatest wetland functions and values, along with optimum wildlife habitat, on every acre enrolled in the program. This program offers landowners an opportunity to establish wildlife practices and protection and long-term conservation on agricultural lands previously converted from wetland. The WRP contains both a cost-share for practices designed to restore wetlands drainage and function as well as the opportunity to sell to the USDA a permanent conservation easement prohibiting any future agricultural use. If the easement is a perpetual easement, it may have important tax and estate benefits to the landowner.

Contact your county USDA FSA or NRCS office for information on these and other federal cost share programs for which you may be eligible to participate.

Federal Taxation Issues[1]

Forest landowners are subject to federal and state tax laws and regulations. Forest landowners who grow timber for a profit, whether through timber sales or by investing in capital appreciation of timber, otherwise known as "banking on the stump", should have a basic knowledge of timber taxation as taxes have a significant impact on the profitability of a business or investment. Every forest landowner should understand how to keep adequate records, deduct ordinary operating expenses, recover capital expenditures such as reforestation and site preparation costs, use basis to claim losses and reduce tax on timber sales, and obtain capital gains treatment of timber sales income. There is a plethora of information available through the world wide web—but it is important to make sure the information is current, as tax laws and regulations on timber have been changing rapidly in the past decade.

Recordkeeping for Tax Purposes

At a minimum, every forest landowner should maintain a record of expenditures on his or her forest. The IRS does not require any specific format for record keeping, so it is up to the individual to determine what best fits his or her situation. A simple Tree Farm Journal or a commercial software package for recordkeeping can suffice for most operations. The important thing is to consistently record all transactions and retain receipts in a usable format.

Another necessary record is that of the land basis and timber basis. Basis is the book value of an investment. Not every forest owner will have a basis in timber, but those who do can use the basis to reduce the tax on timber sales or to claim timber losses. When timber is sold, tax is applied to net proceeds—that is, the gross proceeds less sales expenses and less applicable basis. When timber is destroyed by a casualty or other event, the loss is determined by comparing the loss in fair market value or basis, and claiming the lesser of the two as the loss. The process of using basis to offset income or to claim losses is called depletion.

Ordinary Operating Expenses

Forest owners who grow and manage timber with a profit motive are eligible to deduct ordinary and necessary costs of management in the year of occurrence. Typical costs include boundary line maintenance, mid-rotation herbicide application, tools, equipment, travel to the forest for necessary oversight, fees for forestry educational programming, forest commodity group membership fees, subscriptions for forestry periodicals, road maintenance, and many others. Instead of expensing these costs (deducting them outright), some landowners choose to "capitalize" these costs. This means they are added into the timber basis. In most cases, the forest owner is better off to expense management costs during the tax year they are incurred, rather than capitalizing them. Investors, who take forest management deductions as miscellaneous itemized deductions, may not be able to take the deductions if they use the standard deduction. In that case, the expenses should be capitalized and recovered through depletion when the timber is sold or a loss is claimed. Capitalization must be elected by attaching a statement to the tax return. The decision to capitalize management costs is binding on future year's expenses.

Carrying charges, such as property taxes or mortgage interest or certain other timber related costs may be capitalized or expensed, with the decision made on a year to year basis. Again, the decision to capitalize carrying charges must be announced to the IRS by attaching an election statement with the original tax return for the year in which the election is to take effect. The election cannot be made on an amended return. The forest owner is not allowed to capitalize carrying charges in any year the property is productive. Forest land is productive in any year in which income is produced from its use (such as from hunting leases or timber sales).

Passive business owners face special restrictions on the deduction of management expenses and may be forced to carry the deductions forward until passive income is available. Passive loss rules do not apply to investors; only to business owners.

Capital costs (such as pruning) cannot be deducted outright except when special rules apply, such as in the case of the reforestation recovery rules. Instead they are capitalized by recording them in the appropriate basis account and recovered when the asset is disposed of by sale or by loss. IRS Revenue Ruling 2004-62 clarified that the cost of post-establishment fertilization is a deductible management expense and no longer needs to be treated as a capital expense.

Necessary and reasonable costs of equipment, such as computers, trucks, tractors, and four-wheelers, are treated the same as for any other business. IRS publications, such as **Publication 946 "How to Depreciate Property", Publication 535 "Business Expenses", Publication 463 "Travel, Entertainment, Gift, and Car Expenses",** as well as others can provide excellent guidance on claiming legitimate deductions. These publications can be downloaded from *www.irs.gov* or ordered through the IRS Hotline at 800-829-3676.

Reforestation Tax Deduction and Amortization

As previously stated, capital investments, such as investing in a new forest, are generally not expensed. Instead, the initial costs are recorded in a basis account and recovered later by offsetting the applicable basis against gross sales proceeds, along with sales expenses, to determine net taxable gain. The American Jobs Creation Act of 2004 created a new method of recovering reforestation costs. Under these rules, forest owners can currently deduct outright the first $10,000 of qualified reforestation expenses during the 2005 tax year for each qualified timber property. In addition, the owner can amortize (deduct), all reforestation expenses in excess of $10,000 over an 84 month period using a specified formula. Trusts are ineligible for the initial deduction, but they may amortize all of their reforestation expenses using the same formula.

The Gulf Opportunity Zone Act created a temporary increase in the initial deduction for landowners within the areas affected by Hurricanes Katrina, Rita, and Wilma. These landowners may deduct up to $20,000 for the initial deduction. This additional deduction does not apply to publicly traded corporations, real estate investment trusts (REIT'S), and taxpayers that own or lease more than 500 acres of qualified timber property in total, including timberland inside and outside these zones. There are three different zones for purposes of determining the time period over which this provision applies.

Zones Affected	Effective Dates
Gulf Opportunity Zone	On or after August 28, 2005 and before January 1, 2008.
Rita Gulf Opportunity Zone but outside the Gulf Opportunity Zone	On or after September 23, 2005 and before January 1, 2008.
Wilma Gulf Opportunity Zone but outside the Gulf Opportunity Zone and outside the Rita Gulf Opportunity Zone –	After October 23, 2005 and before January 1, 2008

As an example, assume a forest owner spent $14,000 to reforest a cutover tract in 2006. The owner can claim a $10,000 reforestation deduction in 2006. In addition, the owner can amortize the remaining $4,000 (the amortizable basis) over the next eight years (2006–2013) using this formula:

Year 1: ½ of 1/7 of amortizable basis
Years 2–7: /7 of amortizable basis
Year 8: ½ of /7 of amortizable basis

This means that on the owner's 2006 tax return one-half of ($4,000÷7) or $287 is amortized and deducted. For the next 6 years the landowner can deduct ($4,000÷7) or $571, with the remaining $287 deducted the 8th year.

It is important to claim these deductions on the proper form. Forest owners should take the reforestation deduction on the front of **Form 1040** if they are an investor and should write "RFST" and the amount on Line 36, the "adjustments to total income" line. (See Instructions for 2006 **Form 1040,** Line 36.) If the forest holding is a trade or business, the owner should take the reforestation deduction on the "Other expenses" line on **Form 1040, Schedule C** or other appropriate business form. If the forest is part of a farm, the owner should take the reforestation deduction on the "Other expenses" line of **Schedule F, Form 1040.**

The use of the initial deduction and amortization is not automatic. The forest owner must 'elect' the initial deduction and/or amortization, and must do so on a timely filed return. To elect use of the initial deduction and amortization, the forest owner should file **Part IV of Form T: Forest Activities Schedule** or attach an election statement to the tax return. An election statement must include the identifying number or name of each qualified timber property, the type of activity (such as herbicide application, burning, site preparation, planting, seedling expenses, etc.), the number of acres treated and the total expenditures. This election must be made on a timely filed return for the year in which the expenses are incurred. The owner should then transfer any amortized reforestation expenses to Part VI of Form 4562, Depreciation and Amortization. The amortization deduction is then taken with the initial deduction on the forms listed above.

Cost-Share Exclusion

Many forest owners use cost-share to pay for part of their reforestation expenses. There are special tax rules which apply to these cost-share payments. When cost-share assistance under one or more of the Federal or State cost-share programs is received, the agency will issue a **Form 1099,** which must be attached to the tax return. The payment is then reported as income. Under certain circumstances, the forest owner can elect not to pay taxes on the reforestation cost-share payment. He has a choice of paying tax on the cost-share payment and then deducting the total reforestation costs, using the special rules, or he may choose to exclude all or part of the cost-share payment from income and treat only out-of-pocket costs under the special reforestation provisions.

Specific conditions must be met for the cost-share payment to qualify for exclusion. First, the cost-share program has to be approved for exclusion by the IRS. Programs approved for exclusion by the IRS include the cost-share portion of the CRP, FLEP, WRP, EQIP, and WHIP, plus several state programs (check with your state forestry agency for approved programs).

Second, the cost-share payment must have been for reforestation or site preparation. No cost-share payment may be excluded if the practice is an ordinary operating expense that is deductible. For cost-shared practices in this category the only choice is for the taxpayer to treat the payment as taxable income and then deduct the actual cost of the practice.

Third, the amount of exclusion available is determined using set formulas. The formula to determine the maximum exclusion amount per acre is the greater of (a) the present value (PV) of the right to receive $2.50 per acre or (b) the present value of 10 percent of the average income per acre for the past 3 tax years. These formulas require the use of an appropriate discount rate. This can be the individual's alternative rate of return or the local farm credit bank rate as published every six months by the IRS in a revenue ruling.

To use formula (a), divide $2.50 by the discount rate to determine the amount per acre that is excludable. For formula (b), the amount is 10% of the average income over the past three years, which is then divided by the discount rate. The income could be from a timber sale, hay sales, or any other use of the property. If the tract has been harvested within the last 3 years, it is likely that the full amount of the qualifying cost-shares will be excludable.

As an example, consider the case of a forest landowner who harvests $500 per acre from his timberland. He received cost-share payments of $100 per acre for site preparation the following year. The farm credit bank rate for his location is 6%. Under formula (a) he may exclude $41.67/acre of his cost-share payment ($2.50/.06). Under formula (b) he may exclude $277.77/acre {[($500/3) x (.10)]/.06}. The taxpayer selects the greater of the two—formula (b)—which allows the owner to exclude the entire amount of the cost-share payment from taxation.

The owner will have to determine whether it benefits them more to exclude or include qualifying cost-share payments in the income. Some taxpayers benefit more by excluding the payments; others benefit more by including them and making full use of the reforestation deduction and amortization. Either way, all cost-share payments received must be reported as income. To make the election to exclude the payments, the forest owner must attach a statement to the return stating specifically what cost-share payments they received and formulas and assumptions used to justify exclusion. A reference should be made to IRS Code Section 126 so that the IRS agent has easy access to the regulations allowing this exclusion.

Conservation and Wetland Reserve Programs

Landowners who have enrolled land in the Conservation Reserve Program receive annual rental payments. Currently, these payments are reported as income and taxed. Under Revenue Ruling 60-32, CRP participants who were not actively farming or who did not materially participate in farming activities were allowed to exclude CRP rental payments from the Self-Employment Contributions Act (SECA). Currently, the IRS is proposing to retire Revenue Ruling 60-32 to make all cost-share rental payments subject to self-employment tax, including those to landowners who are not active in farming or those who hire out the required CRP maintenance activities. Check with your tax advisor to see what rules currently apply as no final decision on this question has been made at the time of this publication.

Special Recovery Rules for CRP Expenses

Farmers whose property is in the CRP program may treat expenditures for soil and water conservation on farmland, including reforestation expenses, as expenses in the year incurred. This way they can avoid capitalizing them or treating them under the special reforestation rules. However, the amount deductible in any year shall not exceed 25 percent of the gross income from farming.

Casualty and Other Losses

Timber investment contains an element of risk from losses due to both natural and human caused events. Normal losses, such as the death of seedlings from poor planting practices are not deductible. Casualty and non-casualty business losses may be deductible if the forest owner has basis in the timber. A casualty loss must result from some event that (1) is identifiable, (2) damaged the timber beyond use, and (3) was sudden, unexpected and unusual in nature. Examples include wildfire and storms. The claim for casualty or other losses can be no more than the adjusted basis[2] minus any insurance or other compensation. A 1999 Revenue Ruling identified the depletion block—the unit that is used to keep track of the adjusted basis of the affected timber—as the appropriate measure of the "single identifiable property damaged or destroyed" in calculating a casualty loss deduction. The IRS has issued Revenue Rulings on southern pine beetle losses in timber stands, drought losses of planted seedlings, and casualty loss deductions. The IRS ruled that beetle and drought losses generally do not qualify for a casualty loss deduction because they are not sudden. They may, however, qualify for a business- or investment-loss deduction. Deduction of losses is quite complicated and usually requires the use of a consultant forester to determine the loss in fair market value. The IRS has a Timber Casualty Loss Audit Techniques Guide, available at *www.irs.gov,* which clearly illustrates the determination of the casualty loss deduction.

When a loss occurs, the forest owner is expected to attempt to salvage the damaged timber. If this is possible, the salvage sale may result in a net gain. The forest owner may avoid paying taxes on this gain by claiming an involuntary conversion and reinvesting in qualified replacement property.

Capital Gains and Self-employment Taxes

Timber sales are usually treated as a capital gain or loss. Long-term capital gains qualify for a lower capital gains tax rate. Short-term capital gains are taxed at ordinary income rates. For all timber sales, whether they are considered as long-term capital gain, short-term capital gain or ordinary income, the tax is applied to the net proceeds, not to the gross proceeds. The net proceeds are the gross proceeds less sales expenses and less adjusted basis.

Basis is the investment value of the asset. An asset has an initial basis which is adjusted by depleting the basis when the asset is sold or a loss is claimed. Basis is also adjusted when expenses are added to basis rather than being deducted—a process known as capitalization.

Initial basis varies according to how a property is acquired. For inherited property, the initial basis is "stepped up" to fair market value as of the time of the decedent's death or alternative evaluation date. For gifted property, the initial basis is the basis of the giver and is called a "carryover" basis. If gift tax is paid on the gift, that affects the gifted asset's basis. For purchased property, the initial basis is the proportion of the total acquisition costs as allocated to the asset. The acquisition cost is the purchase price plus the associated costs such as lawyer's fees, filing fees, timber cruising fees, realtor's fee, etc. Below are examples to illustrate the three initial basis determination methods and how they affect the net gain on timber sales.

Inherited Timber
Jane inherits 160 acres of timberland from her father in 1999. A timber consultant evaluates the timber as containing 640 MBF of pine sawtimber worth $224,000 if it were immediately liquidated. This value is reported in the settlement of her father's estate. Jane's initial basis is 640 MBF at $224,000. She sells the timber two years later for $250,000 and incurs $17,500 in consulting fees for handling the sale. Her taxable capital gain is $8,500 ($250,000 less $224,000 of basis and $17,500 sales expenses).

Gifted Timber
Mark has 100 acres of hardwood, which he inherited from his father in 1970. At that time, the timber was valued at $10,000 (a stepped-up basis as in the above example). He depleted $5,000 of the basis when half the volume was harvested in 1960. He currently has $5,000 in timber basis. He deeds the property to his daughter, Sandra and the basis carries over so that Sandra has a carryover basis of $5,000 in the timber. Sandra sells all the timber for $70,000. Her net taxable gain is $60,000 ($70,000 less $5,000 in carryover basis and $5,000 sales expenses incurred as consultant fees).

Purchased Timber
Anthony buys 120 acres of pine plantation and pays $180,000, including purchase price, attorney's fees, realtor fees, and surveying costs. He determines that the land is worth 50% of the total value and the timber is worth 50%. His timber basis is 50% of the total acquisition cost or $90,000. He harvests the timber five years later for a sales price of $110,000. His net taxable gain is $13,000 ($110,000 less $90,000 in basis and $7,000 in consulting fee expenses).

Many individuals do not have proper accounting records for basis. If the timber has not been harvested and there are adequate records, a consulting forester may be able to establish a basis retroactively. The IRS "Timber Casualty Loss Audit Techniques Guide" contains instructions on retroactive basis establishment.

Not all timber sales are treated as capital gains. If the timber sale income is reported as ordinary income, it is taxed at the ordinary rates—a much higher tax rate than the capital gains rate. Another important advantage of treating timber sale income as a capital gain it that capital gains are not subject to the self-employment tax, which is applied in addition to the ordinary income rates. The self-employment tax rate is 15.3 percent for self-employment income of $400 or more. The rate consists of a 12.4 percent component for Social Security and a 2.9 percent component for Medicare. For 2007, the maximum income subject to the Social Security component of the tax rate is $97,500 while the Medicare component is unlimited. The $97,500 applies to wages subject to Social Security or Railroad Retirement tax received during the tax year as well as ordinary income from the timber operation or other ordinary income.

To qualify for long-term capital gains treatment, the asset must have been held longer than 12 months[3]. The maximum long-term capital gains rate for 2007 is 15%. For taxpayers in the 10% or 15% ordinary income tax brackets, the long-term capital gain rate is 5% for the portion of the capital gain, which if added to ordinary income would be taxed at the 10 or 15% tax rates. The mix of 5% and 15% capital gains rates for taxpayers in the 10% or 15% tax bracket is computed when the **Schedule D** worksheet is completed.

The capital gains rates are subject to change and are a constant topic of discussion in Congress. In 2008 to 2010, the special 5% rate drops to 0%, while the 15% remains in place for income above the lowest two tax brackets. In 2011, the rates will rise to 10% for taxpayers in the lowest two tax brackets and 20% for all other taxpayers.

State Income Taxes

For states with income tax, usually the state deductions "piggy-back" on the federal return. This means the same deductions are taken on the state tax return as on the federal return. Even in piggy-back systems, there are often exceptions for specific deductions, such as bonus depreciation. Some states have special forestry provisions. For example, Mississippi has a 50% tax credit that can be applied against qualified reforestation expenses. Check with the state tax agency or local Extension forester to see what rules apply to forest owners.

Forestry Present-Use Valuation Property Tax Relief

Most Southern states offer some type of property tax relief for woodland owners. Qualifying forest landowners in many states can receive, upon approval of application, property tax relief for managed timberland. The amount of tax relief varies widely from state to state; however it is often substantially lower than the tax on non-conservation/farm/forestry use parcels. In some states for example, forestland qualifies if it is:

- individually owned, including certain types of corporations
- soundly managed for forestry, farm or conservation use

Minimum acreage often applies (10 or 20 acres in most states). Application must be made to the county tax supervisor during the regular listing period. Contact your county tax assessor for the prevailing county or state forestry present-use rates and eligibility requirements.

Financial Analysis of a Reforestation Investment

The timber market in the Southern United States is undergoing a dramatic change. Global competition and trends along with a host of other factors have had a negative affect on the profitability of forestland investments. Nevertheless, Southern forests can remain a profitable venture if the right markets continue to exist and the forest owner is prudent in his or her decision- making. Often forest owners can decide if a forestry investment is profitable by using financial analysis tools and procedures used to evaluate other types of investment. Again, the forest owner is advised to obtain professional assistance from tax, finance and forestry experts.

Information Needed to Analyze Reforestation Investments

Like any investment decision, the results of the analysis are dependent on the information that is available and the assumptions that are made. Care should be taken to adequately research the important aspects of any reforestation investment. In most cases, it is useful to calculate several scenarios based on expert opinions of costs, revenues, productivity, market trends, and the like. These calculations based on several different assumptions give a range of results and is called sensitivity analysis. For example, one scenario may be based on current prices, another on expected real price appreciation of timber, another on real timber price declines, another on above average growth using fertilizer and another based on below average growth from drought or insects. This will give a truer picture of the possible risks and returns.

The following information is needed for any basic forestry investment analysis. In some cases additional information may be required or useful for more accurate predictions.

Baseline information needed:

- The productive capability of the area to be planted. A forester can estimate the site index from which future timber yields can be predicted (see Chapter Five: Forest Soils and Site Index). Yields will vary according to the number of trees per acre, frequency and intensity of thinnings, silvicultural activities and intensity, and the expected or desired rotation age.
- The costs of site preparation and reforestation (see Chapter Six: Southern Pine Reforestation). Reforestation costs vary depending on the condition of the site to be reforested. Costs to prepare the site may range from $0 to over $300 per acre. If the site is free of competing vegetation, no preparation may be necessary. Such is occasionally the case if the area has been recently completely harvested and all slash has been removed. Site preparation costs increase to the upper limit as the amount of competing vegetation increases and/or draining, ditching, bedding or fertilization are required. Little or no cost may be required if natural seeding has adequately restocked the area. *Forest Landowner* magazine (archived copies are available at *www.forestlandowners.com*), published by the Forest Landowners Association, has several articles on costs. These numbers, however, are ballpark figures. Contact a local forester or vendor for more accurate information that reflects the facts and circumstances of the particular tract.

- The amount and frequency of management costs required to maintain and protect a vigorous stand. Landowners periodically may encounter such costs as prescribed burning, boundary line maintenance, fire line construction and maintenance, and insect or disease protection. Natural or uneven aged management costs might be herbicide application, prescribed burning, pre-commercial thinning, and disease or insect control.
- Operating expenses such as property taxes, mortgage interest expense, and insurance. Occasional expenses such as surveying costs and management plan preparation can impact profitability. Purchase of equipment, such as a tractor or other reasonable, necessary equipment may be legitimate costs. Miscellaneous costs such as travel to the tract for necessary oversight, attendance at forestry short courses, membership in forestry commodity groups, and subscriptions to forestry publications should be included.
- An estimate of the future value of harvested timber products. Current prices should be used for the analysis since future prices are unknown. If you want to consider the possibility of real price appreciation, increase these prices during the sensitivity analysis. This is done by increasing the present timber price to reflect the expected increase in timber prices over time and then multiplying the result by the expected yields per acre. An expert in timber price projections should be consulted if one is interested in this. Companies such as Forest2Market *(http://www.forest2market.com)* and TimberMart South *(www.tmart-south.com)* are two price reporting agencies that can assist with this service.
- Analysis of investments can be done with or without consideration of inflation. Whether inflation is considered or not, the most important thing is to be consistent. If inflation is included, make sure it is included in all costs and expected returns. According to Bullard and Straka (1998), "whether you include or exclude inflation won't impact the outcome of your analysis unless the investment has costs that are 'capitalized' for tax purposes."
- Cost–Share and CRP or WRP rental payments can be an important part of the total profitability of the forest investment. Cost-share is often limited by availability, acres, and practices. Check with your local cost-share provider to determine if cost-share is likely to be available, the percentage of costs covered, upper cost limitations and upper acreage limits. Conservation Reserve Program, Wetlands Reserve Program, and the new Emergency Forestry Conservation Reserve Program can provide significant increases in forest profitability.
- Hunting leases, pine straw production, cattle grazing, fee-recreation, and other non-timber income and costs should be included in the analysis if they are anticipated as part of the forest operation.

- Financial analysis may be performed before-tax or after-tax. Costs may be reduced by the marginal tax rate since they are deductible. To find your marginal tax rate, search the IRS website for the current year's federal tax rate. Select the rate that applies to the last dollar of your adjusted gross income. For long-term capital gains, such as timber sales, reduce the expected returns by the current capital gains rate (15% for 2007). Be sure to include state taxes in the analysis as well, if the analysis is done on an after-tax basis.

The Texas Forest Service has a Timberland Decision Support System available online that can be used to make cursory investment analysis of loblolly pine plantations in the southern United States. This model is accessible through the Texas Forest Service Web site *(http://tfsfrd.tamu.edu/tdss/default.htm)*. Check with the state or local county Cooperative Extension Specialist to see if there is an available model for timber investment analysis for desired timber species in your area.

Risk of Timber Investment

Growing timber should not be considered a no-risk investment. While timber has been long considered a low risk investment in relation to other investments, the investor still must be aware that there is a risk in any investment. Investments in timber are subject to natural hazards such as tornados, hurricanes, wildfires, ice storms, droughts, insects, and diseases. Good management may provide some risk amelioration, but cannot protect against nature's uncertainties or man-made calamities. The government does not insure timber owners against losses from natural or other events, although private insurance may be available. There is also the risk of declining timber markets, increasing restrictions on land use and harvesting, and substitution of other goods for timber products.

Landowners must also consider the risk of forest investment in terms of their overall investment portfolio and balance the risk of owning timber against their other investments. The old saying is, "Don't put all your eggs in one basket." This is as true for landowners as it is for the person who invests all his money in one stock. If the landowner cannot afford to lose the value represented in the timber, it may be wise to divest of timber and invest in a more secure investment.

Summary

Timber has traditionally been a very good investment for many private landowners. There are various factors which affect the profitability of timber investment. Government cost-share and rental programs provide incentives for private forest management and increase returns to the landowner. Tax incentives reduce the cost of investing in timber, especially for plantation forestry. Financial analysis can help an individual decide if investing in timber is a profitable investment. The returns expected should be compared with other investments in terms of return and risk. Generally, forestry can provide good returns to individuals who are prudent investors.

A consulting forester, an Extension forester, or a state agency forester can help evaluate the potential economic returns for a given tract if the basic information is given.

Suggested Resources

Readings

- Bardon, R.E. and R.A. Hamilton 2005. "Financial Incentives for Forest Management." Publication WON-4. North Carolina Cooperative Extension Service, North Carolina State University. Raleigh, NC. Accessible online at: *http://www.ces.ncsu.edu/nreos/forest/economicspubs.htm*
- Bullard, S.H, and T. J. Straka. 1998. Basic Concepts in Forest Valuation and Investment Analysis. Edition 2.1. Preceda. Auburn, AL. Available through Forestry Suppliers.
- Coder, K. and C. Dangerfield. 1990. "Estimating Returns from Forest Investments." Forest Resources Unit Bulletin 1030. Georgia Cooperative Extension Service, University of Georgia. Athens, GA.
- Dicke, Stephen G. and D.A. Gaddis. 2006. The Basics of Basis. Revised MSU-ES Publication 1983. MSU Extension Service. Missississippi State, MS.
- Gaddis. D.A. and S.G. Dicke. 2006. "Timber Tax Overview. Forestry Income Tax Series." Revised 2006. Publication 2307. MSU Extension Service. Mississippi State, MS. Accessible online at *http://msucares.com/pubs/index.html.*
- Gaddis, D.A. and S.G. Dicke. 2006. Frequently asked questions about timber casualty losses. MTN-M2 Revised. Mississippi State University Extension Service. Mississippi State, MS. Accessible online at *http://msucares.com/forestry/tax/timber_loss_faqs.pdf.*
- Gunter, J. E. and H. L. Haney, Jr. 1984. "Essentials of Forestry Investment Analysis." First Ed. Corvallis, OR. Oregon State University Bookstores.
- Hamilton, R. A. 1997. "Reforestation as an Investment: Does it Pay?" Publication WON-8. North Carolina Cooperative Extension Service, North Carolina State University. Raleigh, NC. Accessible online at: *http://www.ces.ncsu.edu/nreos/forest/economicspubs.htm*
- Hamilton, R.A. and G. van der Hoeven. 2006. "Federal Income Taxes for Timber Growers." North Carolina Cooperative Extension Service, North Carolina State University. Raleigh, NC. Accessible online at: *http://www.ces.ncsu.edu/nreos/forest/economicspubs.htm*
- Haney, H. L, Jr,, et. al. 2005. "Federal Income Tax on Timber: A Key to Your Most Frequently Asked Questions." USDA Forest Service. R8-TP 34. Accessible online at: *http://www.fs.fed.us/spf/coop/library/timbertax.pdf*
- Haney, H.L. et al. 2001. "Forest Landowners Guide to the Federal Income Tax." USDA Agriculture Handbook 718. Accessible online at: *http://www.srs.fs.usda.gov/pubs/misc/ah_718.pdf* or available for purchase through the Government Printing Office at: http://bookstore.gpo.gov (enter 001-000-04693-4 as the Sales Product Catalog).[4]
- Hubbard, W., R. Abt., M. Duryea, and M. Jacobson. 1998. "Estimating the Profitability of your Non-Timber Forestland Enterprises (revised)." University of Florida, Extension Circular 836.

- Larson, Kievell. "Forestry Incentive Programs and Welfare Mentality." Forest Landowner Magazine, Vol 65, No. 6, Nov/Dec 2006, Pgs 17–18. Accessible online at: *http://www.forestlandowners.com*
- Straka T., Greene, J., Daniels, S., Jacobson, M., and Kilgore, M. "Sources for Forestry Incentive Programs." Forest Landowner Magazine. Vol 65, No. 6, Nov/Dec 2006, Pgs 15–16. Accessible online at: *http://www.forestlandowners.com*
- Wang, L. and J.L. Greene. 2006. "Tax Tips for Forest Landowners for the 2006 Tax Year." USDA Forest Service. Management Bulletin R8-MB 128. Accessible online at: *http://www.na.fs.fed.us/pubs/taxtips/taxtips.pdf*

Web Sites

- www.timbertax.org
  National Web site with up-to-date timber taxation issues. Includes federal income tax, estate taxation and other information.

Review Questions

1. What is a cost-share payment and how does it apply to forest management?

2. Name the various federal cost share programs discussed in this chapter.

3. Explain the initial deduction for reforestation and amortization deduction

4. What is a long-term capital gain?

5. How do you define present or current use?

6. What kinds of information are needed to perform a financial analysis of a forestry investment?

7. What kinds of assumptions must you make to perform a financial analysis?

Suggested Activities

1. Use a current spreadsheet program or canned forestry investment analysis software package to conduct a forestry investment analysis of your property. Try a few different scenarios with different inputs, projected market values and projected dates. Examples of online software packages that can be accessed or downloaded free of charge can be found at *www.cfr.msstate.edu/forval/index.html*, *www.fs.fed.us/emc/nris/hd/qsilver/*, and *http://tfsfrd.tamu.edu/tdss/default.htm.*
2. Find out information on cost-share programs available in your state. Make sure you ask about both federal and state programs. You can get this information from your local county/parish forester, FSA, Extension agent or NRCS representative or online at a variety of places. For example, those found at *www.srs.fs.usda.gov/econ/data/forestincentives/.*
3. Contact your local tax assessor's office and ask them what the local conservation use/present use rates are for the county you own timber in. Do some research on how this compares with other counties in your state. Other states in your region. How about nationally? How would you find this information?

Contact your state Land Grant University to see when their next Extension or landowner program is scheduled. Attend the session.

End Notes

1. This section was adapted from USDA Forest Service Management Bulletin R8-MB 126 "Tax Tips for Forest Landowners for the 2005 Tax Year'", by Larry Bishop and John Greene.
2. Basis is the investment value of an assett.
3. For inherited timber, there is no required holding period. For gifted timber, ownership of both donor and recipient count toward the year's ownership.
4. This publication is currently under revision. There have been significant changes in tax law since it was published, including those applying to reforestation expenses, timber sales income, and capital gains rates.

Unit IV:
Intermediate Stand Management for Timber Production

As your forests continue to grow and mature, you will need to carry out certain activities to maintain the health and productivity of your woodlands. These practices are aimed at improving the quality of your future forest products, increasing the growth rate of high value timber, and protecting your forests from harmful pests, diseases, and wildfires. These intermediate management actions usually begin after your forests have become established but before final harvest and regeneration.

The management of your existing forest for timber production will vary according to several factors. It may range from very intensive efforts aimed at maximizing merchantable timber production in the least amount of time, to relatively modest efforts that are primarily structured to just protect your timber investment. The intensity of your woodland management will depend upon existing stand conditions, soil productivity, availability of money and services, the degree of risk you wish to assume, potential value of timber products, and your overall objectives.

Each chapter of this unit will cover the different intermediate management practices that you might utilize in managing your timber. The chapters are separated into management practices for pine and hardwood forests as well as the types of protection that are needed to sustain these timberlands.

Chapter 9:

Prevention and Control of Major Forest Insects and Diseases in the Southeast

If you produce income from forestland or simply enjoy its beauty, wildlife, and other amenities, the health of your forest is important. Naturally occurring insects and disease can have serious, even catastrophic impacts. In recent years, exotic insects, diseases and invasive plants have hurt forest health. The dangers of wildfire can compound damage from storms and ice. The number of potential problems affecting health can discourage forest landowners. However, there is good news! Simply practicing good forest management promotes forest health. But prevention is key.

There are many ways to profitably manage and improve the vigor of your forest while minimizing the risks. Some are as easy as regularly monitoring your forestland, while others involve more complex processes such as timber management. With awareness, most landowners can promote forest health while minimizing risks. As any good doctor knows, prevention is the cheapest and most effective strategy.

What is a Healthy Forest?

Healthy forests contain timber species well suited to the site and provide a range of values to society. But there is no perfect forest. Even in a healthy forest some trees will die due to forest crowding as well as from native insects, disease, wind, ice, and other factors.

Healthy forests benefit the local community through:

- Profit from the sale of timber or other products
- Abundant game and non-game fish and wildlife
- Recreational opportunities
- Aesthetic enjoyment

Healthy forests benefit the greater community through:

- Employment/income from the sale of timber and value-added forest products
- Removal of carbon dioxide from the atmosphere
- Promotion of renewable resources
- Assurance of productive forests for today and tomorrow

Unhealthy forests on the other hand, cause problems to forest owners and society through increased tree mortality, reduced tree growth and vigor and reduced tree quality which leads to:

- Reduced value to wildlife
- Increased risk of catastrophic insect or disease problems
- Increased risk of wildfire
- Inadequate regeneration
- Reduced recreational opportunity
- Increased risk of injury
- Decreased aesthetic enjoyment

Many of Our Forests are Degraded

Many factors have reduced the health of our forests, leaving them in a degraded state:

- Past soil-abusing land management practices
- High-grading of stands
- Excessive logging damage
- Destructive grazing by livestock
- Invasion of exotic plants and pests
- Establishment of tree species not well-suited to the site

Soil-abusing practices include those done during past farming and logging operations where proper practices were unknown or impractical.

High-grading is the practice of harvesting the best quality and largest trees and leaving the smaller and lower quality trees. High-grading is prevalent in stands with mixed species and continues today as "diameter-limit cutting." An example of high-grading would be when a buyer proposes to cut all trees over 14 inches in diameter. When repeated, high-grading results in stands of lower quality and slow growing trees of less valuable timber species and stem quality. These low quality trees make up the "crop" for subsequent generations of forests.

What Can be Done?

Since it is very difficult and expensive to treat insect and disease infestations in a forest, prevention strategies are best. Prevention is *practical* and *cost-effective.* Follow these three prevention M's: **M**onitor, **M**odify, and **M**anage.

- *Monitor* your forest for evidence of insect or disease infestations often. When potential problems are observed, use control efforts to minimize the spread of insects and diseases and/or salvage affected stems.
- *Modify* reforestation decisions (including species selection), site preparation techniques, seedling grading and handling and proper planting techniques to ward off potential insect and disease outbreaks specific to the soils, history and location of the forest.
- *Manage* to sustain a healthy, vigorous forest. Prevent over-stocked and over-mature conditions that can stress forest trees.

Understand Your Forest and Its Potential Health Problems

Understanding forest health requires that you know what resources are present and their condition. These resources include soils, trees, vegetation and wildlife. Understanding site limitations begins with a thorough knowledge of soils, which are the best way to determine a forest's growth productivity and its susceptibility to pests and disease. Some important questions to ask regarding soils are:

- What are the major soil types on my forests?
- What are the conditions of my soils (eroded, compacted, gullied, inundated)?
- Are my timber species well-suited for my soils? (Consider soil limitations and tree condition.)
- What is the growth potential of trees in my soils (referred to as "site index")?

Soil information is readily available through your county (or District) Natural Resources Conservation Service (NRCS) office. They can help you identify and interpret your soil information. A state forestry agency forester or a private consulting forester can assist you in understanding the species suitability and growth potential of your soils.

Additional forest resource information can be obtained by having a resource professional visit your forest. Pertinent forest resources questions include:

- What are my dominant tree species for each forest stand?

- How old are the dominant trees in the forest canopy?
- What are the present stand conditions? (For example, slow growing or crowding, that will increase the risk of insects or disease.)
- Are there any current disease or insect problems?
- Are invasive or exotic plants a potential threat? If so, what are the recommended control measures?

Develop a Forest Management Plan

Proactive forest management promotes forest health and reduces risks. Active management may be required to return your forest to desirable conditions. You will need a management plan for your forest. A written management plan discusses the condition of the forest and outlines how to accomplish your management goals.

As mentioned in previous chapters, a good management plan contains:

- A statement of your goals and objectives for your forestland
- A map or aerial photograph showing property and stand boundaries, access, and important features such as roads and structures
- A forest description, including soil types, site productivity, number of acres, tree species, stand age, stocking, average height, tree condition and health and topographic information
- A timetable of prescribed forest management activities, including thinning, final harvesting, regeneration plans, wildlife practices, and treatment of problems such as insects, disease or invasive plants. A management plan may also include more specific inventory data such as timber volumes and values

Focus Your Plan for Forest Health

A proper management plan should address specific practices that promote forest health and reduce risks. One or more of the following practices may be recommended:

- Regeneration of desirable species suited to soil type while avoiding overstocking (too many trees per acre)
- Thinning to maintain adequate growing space for trees
- Best Management Practices (BMPs) to protect water quality and soil productivity
- Using prescribed fire where feasible to reduce fuel loading and risk from catastrophic wildfire, and to improve habitat
- Removing cull or diseased trees
- Controlling invasive exotic plants that hinder desirable plant and wildlife species

Monitor Frequently!

Forests should be inspected annually or more frequently in times of stress to identify forest health problems. An insect pest such as the southern pine beetle for example, can take over quickly, so quarterly monitoring might be a more ideal inspection schedule in this case. Inspection for specific pests should be done on a schedule that is appropriate to the problem. For example, inspection for storm damage should be done soon after a storm event. Salvage values diminish with time, and damaged or downed trees can serve as hosts for insect pests.

Landowners can access resources from the Internet to become familiar with insect and disease symptoms. Several such sources are listed on the last page of this publication. For a list of some major potential threats to forest health, suggested monitoring schedules and the symptoms, see the Healthy Forests Reference Guide at the end of this chapter.

Use Timber Harvesting Contracts and Monitor All Logging Activities

Poor logging practices can undo years of good management and diminish the condition and potential of forests. Scarred trees can attract insects and disease. Logging debris can increase wildfire risks. Improper use of equipment on fragile soils or wet weather logging can drastically reduce the growth potential of future forests (95 percent or more of the forest's root system is in the top 12–18 inches of the soil surface, the very area impacted during logging).

To reduce site degradation and potential water quality problems from poor logging practices, employ a professional forester to sell your timber. Always use a timber sale contract. The contract should contain:

- The location of your property and the listed (legal) owners
- The description of sale boundaries and how they are marked
- Timber to be cut or left standing and how it is marked
- Location and use of access roads and loading areas
- Method of payment
- Type of harvest (clearcut or a partial cut)
- Best Management Practices to protect water quality
- State, federal, or local environmental or zoning regulations
- Special site protection provisions such as ceasing logging during wet weather
- Length of cutting contract and compensation
- Deck location and road layout (optional but valuable wherever possible)
- Condition of property after logging is complete (roads, fences, culverts, etc.)
- Reseeding of new roads and decks at the completion of harvest (optional)

The contract must pass responsibility to the buyer/logger for excessive damage. Those parties must be held financially responsible and be required to mitigate (fix) the damage they have caused.

Monitoring the logging job is absolutely crucial to avoiding damage to your land and residual trees. Hire a professional forester to act as your agent and include provisions in the contract for monitoring logging jobs *at least daily* during logging. Insist on a pre-harvest plan to avoid problems before harvesting begins. You may monitor logging yourself and raise any questions that you have with the

Figure 9.1 "Poor logging practices can undo years of good management and reduce the condition and potential of your forests. Damage can include compacted and rutted soils and scarred trees."

Credit: Photo courtesy of Dr. James McGraw

logger, buyer (if different from the logger), and the professional forester before problems arise.

Prevention and Control of Major Forest Insects and Diseases in the Southeast

Specific management strategies for common major insect and disease problems are addressed in the following discussion. Many more insects and diseases impact our forests than those discussed below. Those with no preventive practical or cost-effective management solution have been omitted.

Hardwood Insects and Diseases

Hardwood Insects

There are many insects of hardwood trees. Some eat twigs and others eat leaves. A few bore into the stem of trees. Occasionally total defoliation occurs, but rarely do hardwood insects do permanent damage to the forest until natural population cycles build back to significant levels. Most are periodic in nature. They may be at epidemic levels for only a year followed by years with no major problems. There is really little that can be done to prevent or control the episodes of insect defoliation in established hardwood forests. Close observation of the forest and salvage of severely damaged or dead timber is usually recommended.

Hardwood Leaf, Twig and Root Diseases

Leaf and twig diseases of hardwoods are prevalent throughout the Southeast, but like most hardwood insects, are minor and periodic, and treatment is rarely cost-effective. These diseases rarely kill trees, although their unsightly appearance creates concern for many forest owners.

Root and butt rots *(Ganoderma lucidus* and other species) are common on a wide range of species including oaks, ashes, sweetgum, maples and hackberry. Trees decline rapidly, leaves are dwarfed, twig growth slows and the foliage is often off-color. Roots become rotten and spongy, and often woody, tough, yellow or red conks (fungus) emerge from the dying roots. Root damage caused by rutting or soil compaction is the chief cause of infection. Your best defense is a thorough timber sale contract, which ensures soil protection.

Hardwood Stem Diseases

The most serious problem in hardwood stands occurs when the stems are degraded by rot fungi. Rot fungi usually enter the tree through a wound.

Figures 9.2 and 9.3 "Root and butt rots of hardwoods can result from compacted soil and root damage from poor logging practices. Here conks from a Ganoderma butt and root rot are visible."

Figure 9.2 Credit: Edward L. Barnard, Florida Department of Agriculture and Consumer Services, www.forestryimages.org

Figure 9.3 Credit: Photo courtesy NC Division of Forest Resources

Heart rots *(Hericum, Pleurotus, Polyporous* and *Laetiporous* species) are the single most important disease type of southern hardwoods. Fire and logging damage often causes a wound at the base of the tree where the fungus and bacteria colonize and begin rotting the wood in and around the wound. Once the decay begins, there is no control. When practical, you should prevent wounding and salvage (remove) affected stems. If trees not harvested are damaged, a logging contract can be written that requires double payment for damaged trees.

Canker rots *(Polyporous, Poria* and *Irpex* species) invade the stem through wounds and rot the wood. In advanced stages, tree death occurs. Most hardwood species are susceptible, and oaks are commonly affected. Wounds on the tree often attempt to heal, and fruiting bodies (conks) visible on the outside of the stem are symptoms of this rot. Prevent wounding by removing infected stems when it is practical.

Pine Insects and Diseases

Pine insects

Bark beetles are a major cause of pine mortality in the Southeast. Chief among them is the **southern pine beetle** *(Dendroctonus frontalis)*. Minor species such as **Ips** engraver beetles *(Ips* species) and the **black turpentine beetle** *(Dendroctonous terebans)* also are common.

Figures 9.4 and 9.5 "Fruiting bodies on tree trunks may indicate the presence of heart rots or canker rots resulting from tree wounds."

Figure 9.4 Credit: Photo courtesy NC Division of Forest Resources

Figure 9.5 Credit: Eric Carr, California Department of Forestry and Fire Protection, www.forestryimages.org

Southern pine beetles (SPB) are common in all southern pine species and occasionally are found in Eastern white pine. Affected trees exude a resin pitch tube about the size of a popped kernel of corn, the tree tops change from green to yellow and then a rusty red. The adult beetles work just under the bark and construct S-shaped galleries. Stressed trees are most susceptible to SPB, and several factors commonly favor epidemic populations. The stresses that can lead to outbreaks of pine beetle include:

- Drought
- Crowded trees
- Over-mature stands
- Pure pine stands
- Diseased or damaged stands
- Poor soils

Prevent pine beetle outbreaks by promoting healthy, vigorous forests:

- Carefully evaluate soils and match the right pine species to the site
- Thin overstocked stands to promote/sustain vigorous tree growth
- Remove and salvage diseased or damaged stems
- Immediately harvest emerging spots of infestation (include a 75-foot live-tree buffer around the infestation)
- Harvest mature and over-mature stands

Figure 9.6 "A stand with trees infected with the southern pine beetle. The most recently infected trees may have green or only slightly faded crowns. Those with red, brown, or no needles were infected earlier."

Credit: Terry Price, Georgia Forestry Commission, www.forestryimages.org

Figure 9.7 "Pitch tube of different sizes on the tree trunk are often the earliest sign of infection of of bark beetles including southern pine beetles, engraver or "Ips" beetles, and black turpentine beetles."

Credit: Tim Tigner, Virginia Department of Forestry, www.forestryimages.org

Ips beetles are usually attracted to freshly cut, injured or dying trees or slash; the population builds and then they move into surrounding live trees. Evidence of Ips include reddish-brown boring dust in bark crevices and reddish-brown pitch tubes about the size of a dime; galleries under the bark are Y or H-shaped. Color change of the foliage is similar to that of the southern pine beetle. Prevention strategies are similar for all pine beetles. Minimize logging damage to standing trees and promptly remove felled trees to discourage buildup of beetle populations.

Black turpentine beetle attacks freshly cut stumps and the lower trunk of live trees, particularly trees with wounds. Pitch tubes are about the size of a half-dollar. Often this beetle is not fatal, especially in vigorous trees. Promptly remove dead trees and undertake activities which promote stand health and vigor. Minimize root and trunk damage to minimize infestation. This is the only pine beetle that can be economically controlled by using insecticides, especially in yard situations.

Figure 9.8 "Conks on a pine tree indicating the presence of red heart."

Credit: USDA Forest Service Archives, USDA Forest Service, www.forestryimages.org

Figure 9.9 "Long-term infection of red heart in pine can render the tree unusable for lumber."

Credit: Robert L. Anderson, USDA Forest Service, www.forestryimages.org

Pine Stem Diseases

Stem diseases of pines include **red heart** *(Fomes pini)*, **pitch canker** *(Fusarium moniliforme)*, **brown cubical rot** *(Polyporous schweinitzii)* and **fusiform rust** *(Cronartium fusiforme)*. Each of these diseases degrades the main stem or butt of the tree, which reduces its end value. Of these, red heart and fusiform rust are especially common problems.

Red heart is a disease of mature or over-mature pines of all species. The internal rot that it causes renders the tree unfit for use. The affected heartwood becomes light red to brown in color and rotten. The fungus produces a conk that protrudes from the tree. The only practical control is to harvest trees before they become susceptible.

Fusiform rust ranks as one of the most devastating of all pine diseases, occurring on both branches and the main stem. Stem rust galls are the most damaging to tree value. Fusiform rust is very common in loblolly pine and is less common in natural stands than in plantations. Planting eastern white pine, shortleaf pine or longleaf pine on appropriate soils minimizes the problem. Slash pine, which is highly susceptible, should never be planted in susceptible areas. Consider the following strategies to minimize fusiform rust in young stands or regeneration:

Figure 9.10 "Seedlings infected by fusiform should be culled and not planted."

Credit: Photo courtesy of NC Division of Forest Resources

- Plant genetically resistant seedlings
- Plant species on appropriate sites
- Grade seedlings carefully and cull any seedlings with fusiform galls
- Favor rust-free "leave trees" in naturally regenerated stands
- Thin early and selectively to remove infected trees
- Delay fertilization until mid-rotation

To minimize fusiform rust in mid-rotation stands:
- Continue selective thinning to remove infected stems
- Salvage and regenerate following the guidelines for regeneration above if too few non-infected trees are present

Root Diseases

Root diseases cut off the life blood of the tree and may also degrade the quality of the butt log. The major root disease of pine in the South is **annosus root** and **butt rot** *(Heterobasidion annosum)*. Loblolly pines planted on high-hazard sites are very susceptible. High-hazard sites have very well drained sandy topsoils at least 12 inches thick overlaying heavy clays, or also very deep sandy sites which are common throughout the sandhills and scattered throughout the South. The fungus produces a tough, rubbery conk near ground level. The conk is typically tan or brown on top and white underneath and appears from March through early winter. Generally, trees are affected in spots scattered throughout the stand. Symptoms are thinning and yellowing crowns followed by tree death and windthrow. Commonly, infection occurs when spores land on freshly cut stumps or on wounded areas. Then, once infection occurs, the fungus grows through the root system and passes to other live

Figure 9.11 "Multiple branch cankers from fusiform rust"

Credit: Robert L. Anderson, USDA Forest Service, www.forestryimages.org

Figure 9.12 "This butt log of this tree has been severely impacted by fusiform rust."

Credit: USDA Forest Service Archives, USDA Forest Service, www.forestryimages.org

Figure 9.13 "Fruiting bodies from annosus near the ground line of a pine."

Credit: Minnesota Department of Natural Resources Archives, Minnesota Department of Natural Resources, www.forestryimages.org

Figure 9.14 "Treating freshly cut pine stump with borax to prevent annosus infection."

Credit: Photo courtesy of NC Division of Forest Resources

trees through root grafts. Stands often begin to show symptoms between two and three years after thinning.

Prevent root diseases on high-hazard sites:

- Plant longleaf pine rather than loblolly pines on well-drained sandy sites (assuming you are within the natural range of longleaf)
- Plant with wider spacing to eliminate early first thinning
- Thin in the hottest part of the summer since high temperatures are lethal to *annosus* spores
- Treat the surface of freshly cut stumps with borax to kill spores

Figure 9.15 "Pines showing symptoms of annosus root rot."

Credit: Andrew J. Boone, South Carolina Forestry Commission, www.forestryimages.org

Young Pine Insects and Diseases

Young Pine Insects

Young pines can be susceptible to **pales weevil** *(Hylobius pales)*, the **pitch eating weevil** *(Pachylobius picivorous)*, the **deodar weevil** *(Pissodes nemorensis)* and the **pine tip moth** *(Rhyacionia frustrana)*.

Pales weevil *(Hylobius pales)*, the **pitch eating weevil** *(Pachylobius picivorous)*, and the **deodar weevil** *(Pissodes nemorensis)* can be a severe problem in newly regenerated stands. Pales and pitch eating weevils are more common and severe than deodar weevils in newly regenerated southern pine stands. The adult weevils are attracted to the smell of pine resin, such as from freshly cut stumps. On a cutover site, the adults are drawn to the pine stumps, deposit their eggs in the stump and roots attached to the stump where the eggs hatch and the larvae develop. Larvae mature into adults which then move to and feed on the tender bark of seedlings. Weevils kill seedlings by encircling the stem (girdling). Most damage occurs from early spring through mid-summer. In severe cases and with poor planning, total plantations will be lost and must be replanted. Damage can be avoided by following these guidelines:

- Always allow at least nine months, including a summer season, between harvesting and replanting.
- Favor prescribed burning rather than mechanical site preparation. Mechanical disturbance can prolong and enhance the attractiveness of a site to weevils.
- In naturally regenerated areas, you should time harvest cuts during seed fall because newly germinated seedlings are less susceptible to weevils and weevils do not feed during the winter. By the second growing season the weevil populations typically collapse.
- *If none of the above is possible,* plant insecticide-treated seedlings.

Figure 9.16 "Pales weevil on a pine seedling."

Credit: Clemson University—USDA Cooperative Extension Slide Series, www.forestryimages.org

Figure 9.17 "Pine showing regeneration weevil damage."

Credit: Photo courtesy of NC Division of Forest Resources

Pine tip moth is a bud and shoot borer which attacks young seedlings and trees. (Usually the problem goes away once crown closure occurs at 5-6 years of age.) Tip moths retard height growth and deform the tree by repeatedly killing the terminal bud. Most trees outgrow the damage, but this may take several years, especially on poor sites where trees are stressed and slow growing. Tip moths are very common. Low-level populations are always present and evident in newly regenerated pine stands. To prevent pine tip moth:

- Plant longleaf or white pine on suitable sites rather than loblolly or shortleaf pine or use fire and herbicides for site preparation rather than intensive mechanical methods.
- Grow a "messy" stand. Having some grass, weeds or brush intermixed with the pine seedlings reduces the problem (shade favors controlling parasites).
- Treat the terminals of pine seedlings with insecticide after planting if the problem is severe. This treatment is expensive and laborious, but effective.

Young Pine Diseases

The most common disease of young pines is **Brown spot fungus** *(Scirrhia aciola),* a needle disease of longleaf pine that may occur while seedlings are in the grass stage. Infected needles have yellow, amber, or brown spots. Mortality usually occurs only when seedlings are infected for several successive years and needle loss is nearly 100 percent. However, continued partial loss of needles may prolong the grass stage for up to five years, thus allowing less desirable trees species and shrubs to out-compete longleaf for height and growing space. Several measures can effectively reduce the amount of infection.

Figure 9.18 "A Nantucket tip moth."

Credit: Photo courtesy of Dr. James McGraw

Figure 9.19 "Multiple branching and dead branching caused by tip moth damage."

Credit: Photo courtesy of NC Division of Forest Resources

- Plant only high-quality seedlings purchased from nurseries that use fungicides and that grow seedlings at low densities to produce large seedlings.
- Once seedlings are established and the diameter at the ground line is a third of an inch or larger, use prescribed fire during January or February when at least a third of needles are dead in the late fall. Burn each year until seedlings begin height growth.

gure 9.20 "A patch of Chinese privet."

redit: James H. Miller, USDA Forest Service, vw.forestryimages.org

Figure 9.21 "Chinese privet leaves and flowers."

Credit: Ted Bodner, Southern Weed Science Society, www.forestryimages.org

ıvasive Plants

ı addition to insects and diseases, invasive plants can have a major impact on our forests. Invasive lant species vary across the Southeast, but major non-native pests include kudzu *(Pueraria montana),* hinese privet *(Ligustrum sinense),* bamboo grass *(Microstegium vimineum),* Japanese honeysuckle *onicera japonica),* Autumn olive *(Elaeagnus umbellate),* tree of heaven *(Ailanthus altissima),* princess ee *(Paulownia tomentossa),* multiflora rose *(Rosa multiflora),* Chinese wisteria *(Wisteria sinensis),* English y *(Hedera helix),* and oriental bittersweet *(Celastrus orbiculatus).* For more information on lentification and control of invasive plants, see the back page of this chapter. And, always check with our local county extension agent to find out which invasive species are of the most concern in your rea.

Summary

Simply having lots of trees does not make your forest healthy. Past land use, past forest management and harvesting decisions, the presence or absence of natural disturbances and many other factors have shaped your forest. Take a close look at your forest and determine if it is healthy and vigorous and can meet your wildlife, recreation and timber production goals. You and your family have an opportunity to sustain or restore your forest resources. Carefully planned timber harvesting, reforestation and stand tending are usually required.

Many insects and diseases can be prevented or controlled by methods suggested in this chapter. Minimizing damage to stems, thinning stands to maintain vigor, selecting the right species for your soils and proper timing can ensure that your management prevents and minimizes risk.

A stewardship ethic requires that we improve the forest. Our collective efforts can make our forest healthier and leave them in better condition than when we acquired them. You can enjoy a healthy forest. Obtain professional assistance and manage to prevent and minimize loss. (See the included charts to know what to look for, when to look, and how to prevent common pest and disease in the Southeastern forests.)

Suggested Resources

Web Sites

- *http://nps.gov/plants/alien*
- *http://se-eppc.org*
- *www.bugwood.org/weeds/forestexotics.html*

gure 9.22 "*Microstegium* forms dense mats which
but eliminates native low vegetation."

*edit: Ted Bodner, Southern Weed Science Society,
ww.forestryimages.org*

Figure 9.23 "Known by various names including 'Nepalese browntop' and Japanese stiltgrass, *Microstegium vimineum* has become a serious invasive plant, especially in stream bottoms."

Credit: David J. Moorhead, The University of Georgia, www.forestryimages.org

Figure 9.24 "Although Japanese honeysuckle provides benefits for several wildlife species, it can completely cover small trees and eliminate regeneration."

Credit: James H. Miller, USDA Forest Service, www.forestryimages.org

Figure 9.25 "Japanese honeysuckle in flower."

Credit: James R. Allison, Georgia Department of Natural Resources, www.forestryimages.org

Figure 9.26 "Tree of Heaven."

Credit: Chuck Bargeron, The University of Georgia, www.forestryimages.org

gure 9.27 "Tree of heaven can often be seen in
ıtches along road-sides."

*redit: David J. Moorhead, The University of
eorgia, www.forestryimages.org*

gure 9.28 "Autumn olive in flower."

*redit: James H. Miller, USDA Forest Service,
ww.forestryimages.org*

Figure 9.29 "Autumn olive in fruit."

*Credit: Jerry Gibson, Deer Park, AL,
www.forestryimages.org*

Problem or Potential Problem	Where to Look	What to Look For
Logging damage		
Stem damage	Stands being thinned or partially cut	Especially butt log scarring, scars anywhere on stem
Soil compaction, rutting	Throughout logged area, esp. skid trails	Soil compaction, rutting
Weather-related		
Ice	Entire forest	Main stem breakage, fallen trees
Wind	Entire forest, esp. recent thinnings	Main stem breakage, fallen trees
Flooding	Flood-plains, low areas	Immediately confirm extent of flooding, then watch for mortality

Healthy Forests Reference Guide

...hen to Look	Preventative Measures to Reduce Risks	What to Do When Problems are Found
...uring logging operations	Use a Consulting Forester and good sale contract and monitoring	Contact your buyers, consulting forester, loggers to stop logging until problem is resolved
...uring logging operations	Use a Consulting Forester and good sale contract and monitoring	Contact your buyers, consulting forester, loggers to stop logging until problem is resolved
...ithin two weeks	In ice-prone areas, plant local-source pine stock	If main stem breakage is more than scattered, contact a professional forester for advice on salvaging timber
...ithin two weeks	If first pine thinnings are later in rotation, avoid heavy first thinnings Avoid thinning pure stands of Virginia pine In hardwood stands, remove decadent and otherwise cull trees	If main stem breakage is more than scattered, contact a professional forester for advice on salvaging timber
...mediately, then mid-summer ...flooding occurred	Avoid planting upland species in flood-prone areas Remove or breach beaver dams to avoid growing-season flooding	If significant top dieback or mortality eventually occur, contact a professional forester to get advice regarding possible salvage of timber

Problem or Potential Problem	Where to Look	What to Look For
Pine Insects		
Southern pine beetles	Stands with pines 4 inches or larger	Yellow to red crowns and whitish "popcorn looking pitch tubes on lower tree trunk. Especially observe pines struck by lightning and adjacent pine stands.
Ips beetles	Stands with pines 4 inches or larger	Reddish-brown boring dust in bark crevices and reddish-brown pitch tubes (dime size). Fading tree crowns or yellow to red crowns
Black turpentine beetles	Stands with pines 12 inches and larger	Large pitch tubes on lower trunk
Reproduction weevils	Pine stands during 1st year after planting	Seedling mortality. Small irregularly shapec holes in stem. Presence of adult weevils.
Pine tip moth	Young loblolly or shortleaf stands first 5–6 years	Dead or dying tips on branches

Healthy Forests Reference Guide, continued

When to Look	Preventative Measures to Reduce Risks	What to Do When Problems are Found
Quarterly	Carefully evaluate soils and match the right pine species to the site Thin overstocked stands to promote/ sustain vigorous tree growth Harvest mature-overmature stands Consider planting longleaf in its natural range rather than shortleaf or loblolly	Remove or salvage diseased or damaged stems Immediately harvest emerging spots of infestation including a live-tree buffer around the infestation
Quarterly	Preventative measures are similar to those for Southern Pine Beetles Promptly remove felled trees or trees damaged from logging	Remove or salvage infected trees
Annually	Maintain good stand vigor through thinnings and other practices that promote diameter growth such as fertilization. Remove storm or logging damaged stems.	Mortality is infrequent. Salvage only if severe infestation.
February through June	Allow at least 9 months, including a summer season, between harvesting and replanting. If this is done, weevils are unlikely to be a significant problem; or Use prescribed burning rather than mechanical site preparation; or Use insecticide-treated seedlings; or In naturally regenerated areas, time harvest cuts during seedfall.	Replant the following year
Growing season Annually	Plant white pine or longleaf on suitable sites rather than loblolly or shortleaf; or Use fire or herbicides rather than intensive mechanical site prep.	Nothing, but you will get reduced stand growth in early years.

Problem or Potential Problem	Where to Look	What to Look For
Pine Needle Diseases		
Brown spot of longleaf	Longleaf seedlings in the grass stage	Yellow, amber, or brown spots on needles or death or loss of needles
Pine Stem & Branch Diseases		
Red heart on pine	Large mature pines	Conks on tree trunk
Fusiform rust	Pines of any size or age	Stem or branch galls
Pine Root Diseases		
Annosus root rot	Pine stands following thinning	Second year after thinning and later
Hardwood Diseases		
Root and butt rots	Hardwood stands with larger trees	Declining trees with dwarfed and/or off-colored leaves, slow-growing twigs. Roots may be rotten or spongy. There may be conks from the roots.
Canker and heart rots	Hardwood stands pole-size and larger, especially after partial cutting.	Wounds, conks, decay.

Healthy Forests Reference Guide, continued

When to Look	Preventative Measures to Reduce Risks	What to Do When Problems are Found
May through December	Plant high-quality seedlings purchased from nurseries that use fungicides and that grow seedlings at low densities to produce large seedlings.	Use prescribed fire during in January or February when at least a third of the needles have been lost by late fall.
Annually	Harvest trees before they are overmature	Harvest the stand and regenerate
Annually	Plant resistent species on appropriate sites such as shortleaf, white pine, or longleaf pine; or Use prescribed burning twice before thinning on high-hazard sites.	Thin early and selectively to remove diseased trees. Continue to selectively thin.
Annually	Plant longleaf or slash rather than loblolly on well-drained sandy sites. On susceptible sites with loblolly, delay the 1st thinning by using wider spacings. Treat cut stumps with borax.	If infection is severe, clearcut and plant resistent species such as longleaf.
Annually after any partial cuts or salvage.	Good logging contract and monitoring that minimizes soil compaction.	Salvage infected trees.
Annually after any partial cuts or salvage.	Avoid tree scarring of stems.	Salvage infected trees.

Chapter 10:

Intermediate Pine Stand Management

Silvicultural treatments applied in established pine stands are often desirable or necessary to improve the growth rate or survival of the trees. No standard schedule can be suggested for intermediate silvicultural practices, because their application is affected by such factors as stocking, growth rate, site quality, competition and products to be produced from the stand. Common silvicultural activities are release (removal of undesirable hardwoods), precommercial and commercial thinning, prescribed burning, pruning, timber stand improvement and supplemental fertilization. These and other intermediate stand management options may be applied to protect pine woodlands and improve economic returns.

Many management practices, such as thinning, can generate immediate income. But other activities are often quite expensive and are undertaken only when they are expected to increase your forest's future value by an amount greater than the treatment cost, compounded to the time of final harvest. Economic returns generated by stand management depend on local markets, site, weather, stand conditions and costs. The biological objectives and management you introduce to your woodlands are used to increase the amount and value of merchantable material beyond what would be produced without management activities. As a general rule, treatments applied to older stands near harvest and those stands most needing attention result in the greatest economic return.

The following discussions include various practices suitable for intermediate stand management and describe conditions under which these practices are most commonly applied. As you read through this chapter consider a few questions that will help you identify your needs and prospects:

- Do the regenerated areas in your woodland need precommercial or commercial thinning?
- Are there forest health risks present or likely to develop in the near future because of stand conditions, such as overstocking?
- What local market and price situations might affect your opportunities to commercially thin your pine stands?
- Is prescribed burning a consideration of yours, and what are the possibilities of its use in your area?
- What order of priority do you have for management practices in your woodland? (Make a list.)

Release

Southern pines are intolerant of shade and are generally poor competitors with hardwoods that are growing above, beside and below them. Release treatments are aimed at reducing hardwood competition in pine stands. Pine release treatments can be applied shortly after pine stand establishment or during mid-rotation operations. On many sites stands may have been established by natural regeneration or with minimal site preparation that did not effectively reduce existing hardwood competition. As the young pine stands develop, hardwood sprouts from existing root stocks can overtop the pines (Figure 10.1). In this case, pine release involves removing hardwood stems, brush and other heavy vegetation overtopping and interfering with pine seedling growth and survival. Release is normally required within five years after pine stand establishment. Treatment costs for release in this period are not associated with site preparation and establishment costs and can be treated as an expense in the year they occur.

The purpose of early release is to assure that a stand has at least a minimum number of pine trees per acre that are free to grow. To determine the need for release and the likely response of your pine stand to release, you examine the proportion of pine stems to hardwood stems and roughly estimate the basal area per acre for each. (Basal area is the total cross sectional area of all stems, measured at 4.5 feet above the ground and is expressed in square feet per acre.) If the amount of pine basal area is less than that of the competing hardwoods, release will be a helpful management tool. But if pines already compose two-thirds or more of the stand, an early release treatment many not be necessary. In some cases, the competing hardwood stems may have good commercial potential on your woodland site making release efforts costly and unnecessary. In this event, you can simply allow a mixed stand to develop.

Mid-rotation release is, operationally, a more commonly applied treatment. Stands ranging from 10 14 years of age may have a mid-story composed of hardwoods that pose limitations to pine productivity. Despite being overtopped by pines, hardwoods still exert a great capacity to utilize nutrients, moisture and rooting space that could be directed to increasing pine volume production

Figure 10.1 "4-year old pine seedlings with overtopping hardwood sprouts."

Credit: James H. Miller, USDA Forest Service, www.forestryimages.org

Figure 10.2 "Hardwood competition in 13-year old pine stand."

Credit: James H. Miller, USDA Forest Service, www.forestryimages.org

(Figure 10.2). Basal area can be used to judge the need for a release treatment for mid-rotation aged stands. If the hardwood basal area comprises 8 to 10 percent or more of the stand, a positive responsive to release can be expected. In some studies, release of stands from as little as 7 to 8 percent of hardwood has been as effective as fertilization to increase pine volume growth.

Release Methods

It is possible to release individual seedlings and young trees by removing competition from around them with an axe and other hand tools (Figure 10.3). This is a manual practice and can be used if you are willing to devote personal time to the release effort. Labor-induced release generally requires

Figure 10.3 "Brush saws to remove competition around individual pine seedlings."

Credit: James H. Miller, USDA Forest Service, www.forestryimages.org

Figure 10.4 "Directed spray treatment for hardwood control."

Credit: James H. Miller, USDA Forest Service, www.forestryimages.org

follow-up treatments. But most pine release is accomplished by the use of herbicides to kill or suppress overtopping hardwoods to allow the pines to dominate the site. Herbicides can be applied to individual hardwood stems by injection, basal treatment, or foliar sprays. Broadcast treatments can be made by ground, aerial or soil-spot applications with selective herbicides.

Injection or other individual stem treatments are more expensive than broadcast applications for small and numerous stems (Figure 10.4). Herbicides lethal to pines can be used for release if applied as directed sprays or injected into hardwood stems, avoiding contact with pine foliage, or for soil active herbicides, treatment in the root zone. Generally, aerial or ground broadcast treatments with selective herbicides are the common and cost-effective techniques for release (Figure 10.5). Herbicide selection for release treatments must take into account the pine species, seedling age, site conditions (soil texture, drainage) and target species to be controlled. When releasing young stands in an early release, seedling age is an important factor in selecting a suitable broadcast herbicide. Herbicides may have specific age recommendations for treatments in young stands.

Any herbicide you select for vegetation control must be labeled for forestry use in your state. Be sure to read the herbicide label before purchasing, storing, mixing and applying the herbicide. The label includes information about safety, application rate, methods and conditions of application and the effectiveness of the chemical for approved uses. Many landowners contract release treatments to consulting foresters, however, if you wish to do the work yourself you may need a herbicide and/or a pesticide applicator's license. Information on license requirements, training and regulations can be obtained from your university county Extension Office or state Department of Agriculture. Be sure to check with your extension agent, consulting forester, state forester, and chemical company representative in making your herbicide decision. They can answer questions and provide other

Figure 10.5 "Aerial release treatment on ten year old slash pine."

Credit: Ron Halstead, Halstead Forestry & Realty, Inc., www.forestryimages.org

information regarding your particular situation. There are substantial penalties for misuse or application by unlicensed personnel.

The herbicide label contains information on the sensitivity of weed and crop species to the herbicide. Both timing and application rate are critical in herbicide release operations. Improper application may damage crop trees, nontarget vegetation, neighboring crops, result in ineffective weed control, or pose a health hazard to others as well as the applicator. A list of chemicals that have been used for pine release is included in Table 10.1. You must be careful to comply with label specifications and applicable state and federal laws. Herbicides vary in their persistence in the environment and their impact upon soil, water and non-target plants and animals. Know and follow all of your state's Forestry Best Management Practices (BMPs) when planning and making herbicide release treatments. Approved herbicides have been thoroughly tested as part of the registration process. Impact on targeted species, wildlife, groundwater and soil are negligible when used according to the manufacturer's directions. Again, you are encouraged to acquire all the available information possible concerning herbicides and their applications. State and federal agencies are always a good start, especially when your actions may affect much more than just your land.

Thinning

Stand density or stocking (number of trees per acre) is second only to site quality in its effect on yield. Overstocking, or too many trees per acre, leads to reduced quantity and quality at harvest time. Overstocked stands are also more susceptible to forest health problems, particularly pine bark beetle attacks (Figure 10.6). To maintain vigorous crop tree growth and stand health, trees may need to be removed periodically during a stand's life or rotation. These partial harvests are referred to as thinnings. A "precommercial thinning" is used in dense young stands to reduced overstocking by removing nonmerchantable stems. A "commercial thinning" is made when average tree diameters are large enough to harvest merchantable trees. The first commercial thinning in a stand usually produces only pulpwood size trees, but later thinnings may produce chip-n-saw, small sawlogs, poles and other products. Removing poor quality, defective and slower growing trees in a thinning increases diameter growth of remaining crop trees and decreases loss from natural mortality. *Thinning does not increase the total volume or fiber yield of a stand, but it substantially increases the value of the products produced in the stand.* Put another way, thinning helps concentrate the available growth of a

Herbicide Formulation (Brand Name)	Application	Remarks
Hexazinone (Velpar)	Soil spot Broadcast application	Soluble granule or liquid applied in early spring. Carefully check soil texture to determine rate. Age restrictions on applications to young stands.
Triclopyr (Garlon)	Directed spray application Injection or cut surface treatment Broadcast sprays beneath pine canopy	Do not allow contact with pine foliage or desirable vegetation. Not a soil active material. Does not provide grass control.
Imazapyr (Arsenal)	Directed spray Injection or cut surface treatment Broadcast application	Directed sprays on individual hardwood crowns avoiding pine foliage. Over-the-top broadcast treatments possible with 1st year loblolly pine and in mid-rotation stands
Glyphosate (Various generic brands) (Accord Concentrate)	Directed sprays Injection or cut surface treatment Broadcast application with Accord	Directed sprays on individual hardwood crowns avoiding pine foliage. Accord may be applied over-the-top of mid-rotation pines (can be applied with Arsenal in some release treatments)
Metsulfuron methyl (Escort XP)	Broadcast application	Over-the-top applications to established loblolly & slash pines.

This list may not be complete, nor does the inclusion or exclusion of a herbicide constitute a recommendation or endorsement. Check state regulations and label information to determine the suitability and legality of specific herbicide use in your state. Some herbicides are prohibited in certain states, or are labeled only for minor uses, or are labeled only in certain states under local need labels (FIFRA Sec. 24C). Some herbicides or situations may require licensing or certification for purchase or application. Contact you local Extension agent, state forestry agency personnel or consulting foresters for specific recommendations in the state and area where the herbicide is to be used. This listing should be considered as a guide for preliminary planning. Consult labels to be certain a herbicide is appropriate for the intended use and for instructions on application and safety. Follow all state Forestry Best Management Practices for herbicide applications.

Table 10.1 Herbicides for Pine Release

Figure 10.6 "Bark beetle damage in an overstocked slash pine stand."

Credit: Erich G. Vallery, USDA Forest Service, www.forestryimages.org

Figure 10.7 "Precommercial thinning in a an overstocked natural stand of loblolly pine."

Credit: David J. Moorhead, The University of Georgia, www.forestryimages.org

site onto fewer, higher quality trees. Larger diameter trees are more valuable as sawtimber, plywood and veneer than are smaller trees.

Precommercial Thinning

Pine stands established by natural regeneration or direct seeding often have too many seedlings per acre (up to 20,000). Dense overstocking may cause extremely slow growth or stagnation of the developing stand. Stagnation is most likely to occur on poor, dry sites where trees are slow to express dominance and natural thinning is delayed. Consider using precommercial thinning if stocking exceeds 5,000 stems per acre when the stand age is between three to eight years old. Precommercial thinning will reduce density to 500 to 600 stems per acre, resulting in faster growth, healthier stands and greater economic value.

Precommercial thinning can be done by hand using brush saws to leave individual crop trees at spacings of 8 to 10 feet apart. Mechanical precommercial thinning generally involves chopping or mowing 7 to 8 foot wide parallel swaths through the stand, leaving 1 to 3 foot strips of trees between swaths (Figure 10.7). Trees in the strips may be left undisturbed or may be subsequently hand thinned. Chopping swaths in two directions, leaving small groups of trees at 8 to 10 foot spacing creates a checkerboard pattern. The preferred method of precommercial thinning is a combination

ical thinning in parallel swaths followed by hand thinning within the remaining tree rows, 500 to 600 trees per acre.

sure to delay the mowing or chopping operations until the pines are tall enough to develop a ody stem so they are cut below any live needles. While southern pines do not normally sprout, shortleaf is an exception; if the stems have any live foliage after they are mowed or chopped, they will form new shoots and continue growing. Disking is not generally a preferred treatment, because the bareground created can provide a seedbed for additional pine seed germination if mature seed-producing pines are adjacent to the stand. Other treatments such as prescribed fire, herbicides or fertilizers to reduce stocking levels have generally been unsuccessful and are not recommended for precommercial thinning.

Commercial Thinning

Commercial thinning is an intermediate harvest for stocking control, which results in the removal of merchantable trees. It should improve the growth rate and average quality of the remaining trees. It is important that each acre support an adequate number of well-spaced, high-quality, vigorous trees following thinning. The potential value gains from intermediate cuttings are:

- Concentrated growth on fewer, faster growing more valuable crop trees, since each additional inch of diameter adds approximately three to four percent to the board-foot yield at final harvest
- Utilization of trees that would die before final harvest
- Growth of only the highest quality trees to final harvest, eliminating volume growth on low value stems
- Periodic income from the stand over the rotation

The physiology of trees (how they live and grow) affects the timing and intensity of thinning. Southern pines need a lot of light as well as water and nutrients. The crown (upper stem, branches and foliage) of a tree produces the food for wood production and growth according to size and vigor. Overtopped and shaded branches thin out and die, leaving a smaller live crown. Good stem diameter growth requires the live crown be approximately one third of a tree's height. The fastest growing trees are the most successful competitors and remain dominant in a stand. Intermediate and suppressed trees which have crowns below the main level of the stand and even some codominant, with crowns in the main level of the stand will, over time, slow down in growth and may eventually die. This natural thinning process occurs as the stand ages and trees compete for growing space and site resources. A young natural stand having thousands of stems per acre, or a plantation with 600 to 1000 seedlings per acre, will each have fewer than 400 trees remaining by age 40, even without manual thinning as this self-thinning occurs.

Commercial thinning allows you to direct the development of the stand towards the production of more valuable products. It also captures the value of poor quality, crooked, diseased, damaged, limby forked trees that would die over the rotation. Another benefit is to increase stand health by maintaining vigorous growth in order to minimize bark beetle attacks. Keep in mind that the large dominant trees will not benefit substantially from the removal of significantly smaller neighbors, and retaining smaller trees will only benefit if they have large enough crowns. Most forest sites will produce approximately the same total volume growth on fewer good stems as they could on many smaller ones. Your objective of density control by thinning should be *uniform diameter growth* rather than the fast to slow response typical of unthinned stands.

Timing of the first commercial thin depends on stand conditions (maintaining healthy vigorous stand growth) and landowner objectives for the desired rotation and products. In plantation pines, begin to think about a commercial thinning when the trees reach 12–18 years of age, depending upon location. Several stand condition indicators are useful in determining when a stand should be thinned.

- *Live Crown Ratio* of trees in the stand is determined from the length of the live (green branches) crown divided by the total height of the tree. For example, if the average tree height is 45 feet and the average length of the live crown is 16 feet, then the live crown ratio is $(16/45)\times 100=$ 35.5%. To maintain adequate tree growth, live crown ratio should not fall below 33 percent. So, a thinning should be scheduled while trees have adequate live crown ratios.
- *Basal Area* is used by foresters to evaluate stocking or stand density. Basal area (or cross-sectional area) is determined from a tree's diameter at a point 4.5 feet above the ground on the uphill side of the tree. This point is referred to as diameter at breast height (DBH). Diameter at breast height in inches can be estimated by placing a measuring tape around the trunk and dividing the resulting number (the circumference) by pi (B equals approximately 3.14). DBH is more easily measured by the use of a diameter tape which is calibrated to read diameter directly from the measured circumference. This basic equation is used to estimate basal area per tree:

  $$\text{B.A. in square feet/tree} = 0.005454 \times (\text{DBH in inches})^2$$

Basal area per acre is simply the sum of the basal areas of all trees on an acre. When the basal area for southern pines is greater than 100 to 120 square feet per acre then the stand is biologically in need of a thinning (live crown ratios are likely to be approaching 33 percent at these levels). Thinning down to a basal area of 60 to 90 square feet per acre is a common rule of thumb.

Thinning Methods

Several thinning methods can be used once it is determined that a stand is ready to thin. Selection of a thinning method is based on stand origin, density, and uniformity, and owner objectives. Remember, the greatest benefit biologically and economically is to remove poor quality trees to favor high quality crop trees. The following are four common thinning methods:

Row Thinning (in planted pines)
Selected rows are removed from the stand. A row thinning might remove every third, fifth, or seventh row. Note that simply removing rows of trees in the stand will reduce stock but not improve the overall quality of the remaining trees in the stand.

Figure 10.8 "14-year old stand fifth-row thinned with selection in leave rows."

Credit: David J. Moorhead, The University of Georgia, www.forestryimages.org

Selective Thinning (natural or planted pines)
Individual trees are selectively removed from the stand. Tree selection is generally based on position, form, and general health.

Combination Thinning (planted pines)
A combination of both row thinning and selective thinning. This is the most commonly applied thinning method. Removal of rows allows access in the stand for harvesting equipment and the selection thinning in the remaining rows removes poorly formed, damaged, diseased, and low vigor (small live crown ratio) trees to favor high quality crop trees (Figure 10.8).

Strip Thinning (natural pines)
Strips (or narrow corridors of trees) are removed from the stand following the land contours. Selective thinning can be done in the "leave strips" (strips of remaining trees).

The intensity of thinning refers to the number of trees that are to be removed, compared with the number that will remain. Foresters normally decide upon a desired basal area per acre to remain after the thinning. Then the stand may be marked to indicate to the timber harvester that trees that will be cut or those that will remain. In some operations the forester will not mark the stand but provide guidelines and supervise the harvesting crew to ensure that the stand is thinned to the desired specifications. This is most common in first thinnings where pulpwood is the primary product removed. In second thinnings and older natural stands are commonly marked before thinning when chip-n-saw, logs and poles are produced. The number of crop trees per acre and the average distance between crop trees can be determined based on individual tree diameter at breast height. Many recommendations call for approximately 80 square feet of basal area per acre. A simple rule can be applied to help mark your stand.

The "1.75× DBH" Rule calls for the approximate spacing after thinning to be 1.75 times the diameter at breast height and is measured in feet. For example, two 12 inch DBH trees should be separated by 21 feet (1.75× 12=21). Application of this spacing rule results in approximately 80 square feet of basal area per acre regardless of tree size. Table 10.2 provides a guide for the utilization of this rule.

The "1.75 × DBH" Rule does have some inherent limitations. It fails to consider stand age, site quality and environmental conditions. You may obtain better results by using a table prepared by foresters for the particular species you are managing. Table 10.3 is a sample table prepared for loblolly pine. These species-specific tables generally require more basal area per acre to remain in stands on better sites than on lower quality sites. More basal area per acre is normally recommended

DBH (inches)	Spacing (feet)	Trees Per Acre	B.A. Per Acre (square feet)
6	11	400	80
7	12	300	80
8	14	225	80
9	16	175	80
10	17	150	80
12	21	100	80
14	25	70	80

Table 10.2 Pine Thinning Guide using "1.75 × DBH" Rule

Site Index, Base Age 50						
	60	70	80	90	100	110
Age	Basal Area Per Acre (Square Feet)					
15					58	63
20		61	67	72	76	81
25	66	73	78	82	85	89
30	74	80	83	87	90	95
35	79	84	87	89	93	98
40	83	86	89	92	96	100
45	85	88	91	94	97	102
50	87	89	92	95	99	104

Table 10.3 Loblolly Pine Thinning Guide Showing Basal Area Per Acre Remaining for a Range of Ages and Sites

for older stands than younger stands as well. The recommended basal area per acre is usually lower in longleaf pine stands and higher in shortleaf.

Foresters frequently modify recommendations obtained from thinning guides, based on their experience. Local markets and environmental conditions are often indicators of these needed changes. For example, in areas of reputed severe weather conditions resulting in frequent tree damage, foresters advise landowners to retain a higher basal area per acre. In areas notorious for ice and snow damage, landowners may be advised to lower the intensity of thinning, leaving more trees per acre.

The timing of a thinning can also be critical in some situations. For example, if moderate bark beetle activity is occurring in the area, the thinning should be performed in the winter months. If there are high bark beetle outbreaks and activity, then delay the thinning because harvesting activity and damage to residual trees after thinning may attract bark beetles. If the stand is in an area of high *annosus* root rot hazard (well-drained soils with at least 65 percent sand in the first 12 inches and a low seasonal water table), thin during summer months when the high temperatures reduce the viability of the fungal spores. Avoid thinning during cooler wetter months when the spores would be more active, increasing the potential for spreading annosus root rot infection from tree to tree.

Thinning Recommendations

Longleaf Pine

Natural longleaf stands typically start as dense clumps of seedlings that will self-thin resulting in well-pruned dominant stems. If precommercial thinning is desired, reduce young stands to 500 to 600 stems per acre. Once stands reach commercial thinning size, thin lightly every 10 years to reduce the stand basal area from 120 to 80 square feet per acre.

Slash Pine

Slash pine respond favorably to thinning at young ages (12 to14 years old). Because slash pines self-prune readily in dense stands, early thinnings are needed to maintain adequate live crown ratios of 33 percent or more. Delaying thinning to 15 to 20 years of age may result in trees with marginal live crown ratios (below 33 percent) that respond poorly to release. Repeat thinnings on 5 to 10 year intervals, depending on site quality, to residual basal areas of 60 to 90 square feet per acre.

Loblolly Pine

Loblolly is the fastest growing of the southern pines on sites suitable to pine growth. Plantations on good quality sites respond to intensive management starting at age 12, with thinnings as frequently as every five years. Such a schedule can produce sawtimber in considerably less than forty years, although pruning may be required to produce clear sawlogs when stands are thinned to lower densities at young ages. Many growers delay the initial thinning for a few years to allow the trees to lf-prune the first 16 foot log. But the live crown ratio should not be allowed to fall below 33 percent efore the first thinning is applied.

ortleaf Pine

ortleaf is a slower growing pine on most sites compared to loblolly pine. In mixed loblolly-shortleaf ands, shortleaf is often removed in intermediate cuttings. Overstocked pure shortleaf stands are inned less frequently than loblolly, but are still reduced to approximately 80 square feet of basal ea per acre.

n intensive commercial thinning schedule may not always be practical. Volumes available for moval could be too small and prices too low to cover harvesting and hauling costs. This limitation is enerally restricted to small stands, poor sites and remote locations. When conditions do allow mmercial thinning, profitable responses may be obtained from even a single intermediate harvest. hinning should be performed as early in the rotation as practical since growth rate and response ill decline as age increases. Commercial thinning is an opportunity for you as a woodland owner to ceive intermediate profits by improving your stand. If done properly, it will bring you profits now nd in the future.

rescribed Burning

istorically, fire, whether natural, accidental or intentional, has been important to the ecology of our uthern pine forests. Because it affects different plants in different ways, fire can have an important fect on the type of vegetation in a forest. In pine stands where fire has been excluded, low-quality ardwoods often occupy much of the land and make it unsuitable for forestry, wildlife or agricultural ses. Wildfires, of course, can devastate a stand. However, where periodic prescribed fires have been sed in the management of pine stands, the forest floor is open and park-like, annual plants and low-rowing sprouts are plentiful for wildlife food, and there is less fuel to feed a wildfire.

The total impact of different types of fire (wild and prescribed) on the forest environment is not ell understood by many forest landowners or the general public. Consequently, many isconceptions on the use, value, necessity and environmental impact of prescribed fire exist. Table 0.4 gives a comparison of prescribed fire to wildfire to further your understanding of the inherent ifferences.

Indians and early settlers were aware that fire could be used to control brush and hardwooods in ine stands. But the fires they set were not controlled and frequently were devastating. Foresters arefully plan and choose conditions under which they conduct prescribed burning, so that low tensity, controlled fires accomplish some clearly defined management objectives. Prescribed urning is considered to be an indispensable and economical tool for managing much of the pine rest area in the South. Prescribed burning is an important tool for:

- Reducing hazardous fuel accumulation to minimize wildfire risk and damage

- Improving wildlife habitat and browse
- Restoring and managing native plant associations
- Controlling understory hardwoods to increase pine productivity
- Controlling disease (Brownspot and annosus root rot)
- Enhancing appearance and value of a stand
- Improving access within a stand
- Preparing sites for seeding or planting

Developing a written burn plan is an essential part of using prescribed fire. The written plan details the objectives of the burn, stand and fuel conditions, the firing technique, season and weather conditions required to burn, smoke management, personnel and equipment needs, permits and leg requirements, setups for one-day burning blocks, and a means to evaluate the effects of the fire. Overall, the plan serves as a checklist to ensure that all aspects of planning and executing the burn are considered before the fire is actually set.

Prepare the plan before the burning season by visiting each stand to determine site conditions, areas to be protected or that need special treatment, and the need for and placement of fire breaks in order to set up the stands into one-day burning blocks. Use existing barriers such as roads, streams, field borders as fire breaks where possible and add plowed breaks as necessary before the

Feature	Prescribed Fire	Wildfire
Timber	No damage when properly applied	Varies greatly, can destroy all trees
Water	No adverse effect when properly applied	Can contribute to silting and reduce water quality
Soil	Litter layer not completely burned and soil not exposed when burns are conducted when soil is moist	Bare soil often exposed, water runoff increases, erosion may become a problem
Wildlife	Improves cover and conditions for some game and non-game species	Varies greatly but can destroy wildlife, cover and forage
Air	1/3 less smoke/ton of fuel than wildfire, lasts only a few hours, burns conducted when atmospheric conditions favor smoke dispersion	Smoke reduces visibility, creates local hazards to transportation, may last for many days
Recreation	Improves accessibility for hunting and other recreation uses	May destroy the forest for recreational uses

Table 10.4 Environmental Effects of Fire

ırn. The one-day burning blocks allow the areas to be prescribed burned during daylight hours to ʼoid smoke management problems that typically occur when fires burn or smolder at night.

Longleaf pine is quite well adapted to fire. It can even survive prescribed burns while still in the edling or grass stage that would be lethal to other southern pine species. Brownspot needle blight ın pose serious regeneration problems for longleaf pine, and prescribed fire is an effective way to ntrol the pathogen. While longleaf seedlings in the grass stage are quite tolerant to prescribed fire, e caution when seedlings enter their characteristic rapid height growth stage when they come out the grass stage, as hot fires can damage the newly elongated shoots.

The other southern pine species are not as fire-resistant as longleaf pine, especially while still in the edling stage. However, all southern pines (except white pine) may be put on a schedule for escribed burning when trees are more than 15 feet tall. Under normal conditions, this height will cur when the trees reach approximately eight to ten years of age. The initial burn is the most fficult and hazardous to conduct, requiring cooler temperatures and steady wind to dissipate the eat to avoid scorching the crowns. Usually the first burn to reduce fuel buildup is conducted in ınter periods, when conditions are less volatile. Prescribed fires should then be repeated at pproximately two to five year intervals, depending upon the rate of litter and brush accumulation. riodic fires in general will inhibit some root rot diseases.

If hardwood understory control is one of your objectives, you should follow the initial winter burn ith one or more growing season burns (Figure 10.9). Growing season burns, following hardwood afout, can effectively top-kill sprouts and further weaken hardwood roots stocks by killing the food oducing foliage. Most thinner-barked hardwoods that are less than three to four inches in diameter e susceptible to top-kill by growing season burns.

gure 10.9 "Backing fire in a 15 year-old stand ıring growing season (early May)"

edit: David J. Moorhead, The University of eorgia, www.forestryimages.org

Fire can also be used for seedbed preparation near the end of an existing rotation. After competing vegetation is controlled, a hot fire can expose enough mineral soil to improve seed germination. For this to be economical and effective, your woodlands must, of course, provide an adequate seed crop. You may be able to anticipate the available seed crop by estimating the number of maturing cones in the trees' crowns.

The difficulty in determining the correct burning conditions (wind, rain, relative humidity, temperature, fuel accumulation and mixture and air stability) and the need for adequate firebreaks requires that forest landowners consult their state forestry agency for advice and assistance. Many laws apply to prescribed burning, and permits are normally required. For example, it is a matter of courtesy and required by law in most states that anyone planning a prescribed fire give all adjacent landowners at least five days written notice of the intention to burn. Because of the technical and legal complexity, and the danger inherent in a prescribed burn, it is very advisable to contract a consulting forester to conduct the burn.

Prescribed burning is well worth the cost per acre when it is done properly. By maintaining firebreaks, subsequent burns become less expensive and easier to conduct.

Pruning

The most common defect in wood marketed as sawtimber is knots caused by branches. Controlling branch growth can therefore be as important to producing value as increasing stem growth. Fortunately, the major southern pines naturally prune themselves when grown in well-stocked stands. Natural pruning proceeds from the ground up, as a result of shading. In poorly stocked stands or widely spaced plantations, natural pruning may be delayed or prevented, producing limby, poorly formed sawtimber. From an economic perspective, no advantage is gained by pruning until virtually all branches on the lowest sawlog (17 feet above the ground) are removed and clear wood covers the wound. Pruning is a labor intensive and expensive operation. However, if you are willing to dedicate personal time to the management of your woodlands, costs can be significantly reduced. [Editor's note: Reviewers commented that in today's market, it is difficult or impossible to recover pruning costs when selling your timber.]

Many factors affect the seriousness of knot defects. Dead branches are slow to detach from the main tree stem and develop into loose or black knots. These black knots are a more severe defect than knots caused by living limbs. Large diameter branches cause large knots in the stem which lowers lumber quality and grade. Note that pruning wounds do not "heal" in the same way that human tissue "heals" after being cut. Trees respond to injury of this nature by developing tissue (callus) that grows over the wound. Wounds created by pruning small diameter live branches create callus more rapidly than those formed by cutting dead branches.

Pruning is an expensive management option, yet it can improve log quality for sawtimber production. Following the established guidelines listed below will result in cost effective managemen and future benefits.

- Prune in several stages beginning when the stems are 3 to 4 inches DBH to minimize the knotty core (Figure 10.10). Start about head high pruning small diameter live limbs. The lowest remaining limbs should be pruned in two or three passes over a period of years until the lowest 17 feet are pruned.
- Remove lower limbs, leaving the top 40 percent of the tree with live crown. This should maximiz the volume growth of clear stem wood.
- Prune only crop trees, those 100 or so per acre of the largest and best trees to be retained to fin sawtimber harvest.

Figure 10.10 "Pruning longleaf bole."

Credit: Bob Farrah, USDA Forest Service, www.forestryimages.org

- Prune close to the bark of the main stem; follow suggested pruning procedures (i.e. do not cut into the branch collar; cut should be smooth and flush, no tearing of bark, etc.).
- Prune live branches in the dormant season, preferably late winter to early spring, using hand, pole saws, or power pruners.
- Prune in stands where density is controlled, such as by thinning, to maximize growth of pruned crop stems.
- Keep good records of pruning so potential buyers may expect unusually high quality clear wood.
- Prescribed burning may also help prune lower limbs, particularly for longleaf pine grown at wide spacings.

mber Stand Improvement (TSI)

imber stand improvement, or TSI, is a term used identify forest management practices which nprove the vigor, stocking, composition, oductivity and quality of forest stands. It is an termediate stand management practice that sults in the removal of cull trees and unwanted ecies. The improvement is accomplished by the moval of these poor trees, allowing crop trees to lly use the growing space. The chief aim of TSI continued production of more and better timber products. TSI practices can be used to convert ixed stands into productive forests of desirable species. It can speed up the growth and improve the uality of the trees in your forest. Undesirable stems are eliminated either chemically or echanically; they may be deadened, felled or removed from the site. Sometimes TSI cuttings can be ld as poles or firewood for campgrounds. Trees you will want to remove during TSI operations clude the following:

- Suppressed trees that will not live until the next thinning
- Trees too crooked, forked or limby to make quality sawlogs
- Trees with fire scars and injuries from insects, disease, wind or ice
- Trees of a species not suited to the site

- Trees that are over-mature or slow growing
- Any tree that will not contribute to the net value of the stand before the next thinning
- "Wolf" trees with excessively large crowns that occupy too much growing space or shade out mor desirable species

Methods of TSI

There are many recommended methods for eliminating undesirable trees. In many cases, TSI is a practice that you can perform yourself to keep costs down and improve your future benefits. The following sections explain individual treatments that are commonly used.

Cut and Remove

Undesirable trees are felled and usually left on the ground. Where economical, use these low quality trees for salable products such as firewood, mulch or fiber.

Girdle

Cut and remove a band of bark all the way around the stem (some species are very difficult to kill in this manner).

Frill

Make a continuous cut around the stem and apply herbicide to the cut.

Basal Spray

Apply herbicide to small stems (less than three inches in diameter) by spraying the bark of the lowe stem. This is usually applied to hardwood trees in the winter for the release of desirable pines.

Spot Treatment

Apply herbicide to individual tree, using a directed gun or spray on the foliage. Spraying in spring o summer is best.

Stump Treatment

Apply herbicide by spray or brush to the fresh stump surface after felling the tree to prevent it from sprouting (usually within 24 hours of cutting).

Injection

Apply herbicide to individual stems using a tree injector or hypo-hatchet. Such tools insert a metere amount of herbicide beneath the bark at intervals around the stem.

ason of year, vigor and size of tree, species and weather are several factors that will determine the fectiveness of a herbicide treatment. Injection of oaks and red maple generally works best in the ll, but other species may be more susceptible in the spring or summer months. Winter injections ay reduce the amount of crown kill.

Contact your county extension agent or local forestry professional for more information related to erbicides that are labeled for TSI and appropriate safety and application information.

ipplemental Fertilization

ertilizers are applied to more than one million acres of forestland annually. Southern pines, articularly slash and loblolly, can respond dramatically to proper fertilization. It is important to nderstand that *improper* fertilization can result in *decreased* growth rates, investment loss, damage and en death of your trees. Response is best when other management practices are being applied and n nutrient deficient sites. Stand age, species, soil characteristics, leaf area, and landowner objectives e important factors in determining the need and potential response to fertilization. Knowledge of our soil type and soil conditions combined with a chemical soil test will help identify sites and stands at will respond to specific fertilizers such as phosphorus. You can obtain more detailed information om sampling pine foliage in the winter to determine nutrient status. Both soil and foliage tests can e done by private laboratories in the South and at many universities.

ertilization at Site Preparation and Planting

hosphorus (P) is the element that is added as a preplant treatment. Typically, very poorly to omewhat poorly drained flatwood sites in the Lower Coastal Plain have the greatest need for replant P additions. Sites with poor surface drainage are normally "bedded," creating raised lanting beds so seedling roots are above the standing water table in the spring. P is generally added ist ahead of the bedding operation. There are also upland sites in the Upper Coastal Plain that are deficient as well. Generally, these are well-drained loamy to clayey soils which have not been in ecent cultivation.

arly Post-Plant Fertilization

ites which are P deficient can be ameliorated with P fertilization after planting. Seedlings will exhibit oarse pale yellow-green foliage. Loblolly pine is more sensitive to low P than slash. A foliar analysis ill confirm the problem.

Most planting sites will have enough nitrogen (N) in the soil nutrient pool to supply the needs of ewly established plantations for several years. Cutover sites in particular will have increased rates of I mineralization from the breakdown of the litter and logging debris and exposure of mineral soil. ontrol of herbaceous weed following planting is a greater priority than supplemental N fertilization.

Percent by weight*			
Name	N	P_2O_5	P
CSP or TSP		45	20
DAP	18	46	20
UREA	46		
Ammonium Nitrate	34		

* N and P_2O_5 are the first and second numbers in the analysis on the fertilizer container; for example, 18-46-0 for DAP means that 100 pounds of fertilizer contains 18 pounds of elemental nitrogen, 46 pounds of P_2O_5 (which is 20 pounds of elemental phosphorus), and zero pounds of potassium (K).

Table 10.5. Common fertilizers and their percent by weight of nitrogen and phosphorous.

Mid-Rotation Fertilization

This is the most common application of N or N + P in forestry. While most sites can supply the initial demand for N, as the stand develops and tree size increases, N can become the factor limiting growth. Generally, at ages 5 to 10 years this occurs, leading to reduced leaf area. If growth is to be accelerated or maximized, stand leaf area must be increased. This is done with N or N +P fertilization. The duration of a growth increase from one N fertilization is 5 to 7 years so the value of the treatment needs to be captured by a harvest at the end of the response period. Likewise, if you fertilize with N at young ages, N will have to be periodically added to maintain the growth response.

Fertilization in conjunction with thinning is a common mid-rotation treatment. N or N + P is added following thinning where crop trees have been selected. The stand is then allowed to grow for 7 to 10 more years and thinned again or harvested to capture the fertilizer growth response. Application is usually achieved by aerial means in early spring or late fall. A significant amount of nitrogen can be taken up by competing hardwoods which can be controlled by various the release treatments discussed earlier. Intermediate and suppressed trees seldom respond to fertilizers. In fact they drop out of the stand more quickly than if no fertilizer is used. Typical rates for loblolly and slash pines call for up to 200 pounds of N and 25 to 40 pounds of elemental P per acre. Longleaf pine has a lower nutrient requirement and N rates are limited to 100 to 125 pounds per acre.

Table 10.5 presents some commonly applied fertilizers for your use in preliminary planning. For further information, contact your county extension agent or consulting forester.

Summary

Considerable information is available about the management of pine stands and the expected performance of those stands. You should rank management options (release, precommercial or commercial thinning, prescribed burning, pruning, timber stand improvement or supplemental fertilization) and select those activities that are compatible with your overall management objectives and limitations. Management practices should be undertaken with an understanding of costs, risks, responses and returns. This information in turn depends on site and stand characteristics as well as external factors affecting markets, including risks, labor, materials and equipment availability. An anticipated schedule of intermediate stand management practices should be carefully incorporated into your overall forest management plan. This will optimize your production of desired timber products at specified intervals in acceptable locations and quantities.

Review Questions

1. What are five management options which may be applied to protect pine woodlands?

2. What are four reasons why timing and application rates are critical in chemical release operations?

3. What is overstocking?

4. Name five methods of herbicide application.

5. Foresters carefully use prescribed burns to economically accomplish several management objectives. What are three benefits resulting from prescribed burns?

Suggested Resources

Readings

- Cantrell, Rick L., George M. Hopper, and George T. Weaver. 1995. "Renewable Resource Notes: Using Site Index To Determine Forest Site Quality." *Publication SP373.* Tennessee Cooperative Extension Service, University of Tennessee. Knoxville, TN.
- Dickens, David, Dave Moorhead, Coleman Danagerfield, and Steve Chapman. 2004. "Thinning pine plantations." *Georgia Forestry Commission Forest Stewardship Publ. www.gfc.state.ga.us/Publications/RuralForestry/PineThinning.pdf*
- Duryea, Mary L. and James C. Edwards. 1987. "Planting Southern Pines." *Circular 767.* Florida Cooperative Extension Service, University of Florida. Gainesville, FL.

- Duryea, Mary L. and John T. Woeste. 1992. "Forest Regeneration Methods: Natural Regeneration, Direct Seeding and Planting." *Circular 759.* Florida Cooperative Extension Service, University of Florida. Gainesville, FL.
- Harrington, Timothy B. 2002. "Silvicultural approaches for thinning southern pines: Method, intensity, and timing." Warnell School of Forest Resources and Georgia Forestry Commission Publ. No. FSP002. *www.gfc.state.ga.us/Publications/RuralForestry/SilviculturalApproaches.pdf*
- Ham, Donald L. 1980. "Fire In The Forest: Good and bad." *Circular 606.* South Carolina Cooperative Extension Service, Clemson University. Clemson, SC.
- Kessler, George D. and Jack B. Cody. 1984. "Forestry as an Investment: How to Compare Pine Trees to Other Investments." *Circular 600.* South Carolina Cooperative Extension Service, Clemson University. Clemson, SC.
- Monaghan, Thomas A. 1988. "Timber Stand Improvement." *Publication 1281.* Mississippi Cooperative Extension Service, Mississippi State University. Starkville, MS.
- Moorhead, David J. 1989. "A Guide to the Care and Planting of Southern Pine." *Management Bulletin R8-MB39.* USDA Forest Service, Southern Region. Atlanta, GA.
- Moorhead, David. 2002. "Fertilizing Pine Plantations: A County Agents' Guide for Making Fertilization Recommendations." Warnell School of Forest Resources, The University of Georgia, Athens, GA.
  www.bugwood.org/fertilization/csoillab.html
- Roth, Frank A. 1991. "Site Preparation Methods for Regenerating Pines." *Publication FSA5002.* Arkansas Cooperative Extension Service, University of Arkansas. Fayetteville, AR.
- Wade, Dale, and James Lunsford. 1988. "A Guide for Prescribed Fire in Southern Forests." United States Department of Agriculture, Forest Service Southern Region, February 1989; *Technical Publication R8-TP 11.*
  www.bugwood.org/pfire/

Web Sites

- www.bugwood.org
  The Bugwood Network
- http://sref.net
  Southern Regional Extension Forestry
- http://sref.info/courses/mtf1/
  Master Tree Farmer I & II

Videos

- "Uneven-Aged Management of Loblolly/Shortleaf Pine." Cooperative Extension Service—Louisiana State University. (A short film covering natural regeneration methods, cutting techniques and mathematical computations that result in a sustained yield from forestlands.)

Chapter 11:
Intermediate Hardwood Stand Management

Despite the attention given pine regeneration and management, nearly two-thirds of the South's commercial forestland is of the hardwood timber type: oak-hickory, oak-pine, oak-gum-cypress, elm-ash-cottonwood, maple-beech-birch, chestnut oak and southern scrub oaks. Furniture manufacturing, pallet production, railroad tie replacements and high quality paper are all industries that have traditionally depended upon the hardwood resource.

Interest in more intensive management of hardwoods is growing. This is because use of hardwoods is outpacing the supply of accessible hardwood volumes of acceptable species. This discrepancy is causing forest owners and industries to increase the intensity of their hardwood management programs that provide for adequate natural and artificial regeneration and intermediate treatments aimed at the production of high value hardwoods. At least two factors prompt optimism about intensive hardwood management:

- Hardwood markets and utilization are improving
- Present stocking and growth rates do not reflect the potential of managed stands

Management of hardwood stands is quite complex. Hardwood stands are typically comprised of a number of species growing together, and the majority of stands have been neglected or mismanaged. Certainly there are some extremely good stands, but the average hardwood stand is *not* fully stocked with desirable trees. Many hardwood stands consist of remnants following one or more partial

Figure 11.1 "Skidding damage common to hardwood stands that have been repeatedly high-graded."

Credit—Jeff Stringer

Figure 11.2 "High quality hardwood production is possible on a wide range of sites in the south. Here veneer white oak is being produced on an upland site."

Credit—Jeff Stringer

harvests, which removed the largest, highest quality and most vigorous hardwood trees. These high-grade harvests have resulted in poorly stocked stands of low value species and poorly formed stems. Unfortunately, high-grading continues to be one of the most common harvest techniques in the South. An example is Kentucky where approximately 70 percent of harvests in the late 1990's were classified as either a high-grade or a diameter limit cut conducted with little regard for future stand development. Not surprisingly, the growth rate of these stands in volume and value is disappointing. Given the situation, the principal alternatives for more intensive hardwood management are:

- Harvest the stand and establish a hardwood plantation where appropriate natural regeneration is lacking
- Perform a harvest cut and plan for natural regeneration of the site
- Rehabilitate the stand through intermediate treatments

ntermediate treatments for hardwood stands are practices that maintain proper stand density thinning), remove overtopping undesirable trees (cleaning), develop high quality trees (release), emove poorly formed commercial or non-commercial species (improvement), or develop advanced egeneration required for future regeneration (mid-story removal). These treatments can be ompleted separately or in some instances combined. Many intermediate treatments are suitable for se in both naturally and artificially regenerated stands and are often required to achieve nanagement goals.

It is also important to understand that intermediate treatments are not capable of fixing all roblems. Regeneration harvests and associated silvicultural activities such as site preparation reatments in conjunction with intermediate treatments are needed to maximize forest growth and evelopment. Typically species composition is greatly affected by regeneration treatments and site reparation activities. These activities have considerable control over species composition, and ntermediate treatments are often used to fine-tune species composition and provide for adequate rowth and development of the species present.

Greater income and more attractive forests can be obtained through the growth of high-quality ardwoods. Incomes from high-quality trees can be six to ten times higher than from low-quality rees. High-quality hardwoods also provide aesthetic values not found in short rotation forests. Many ommercial hardwood tree species thrive in the South's favorable climate and soils. Most species will roduce greater incomes if grown for quality. As you read this chapter consider the following uestions to help you identify the management needs of your woodlands:

- How many of your hardwood acres are on sites best suited to continued hardwood management?
- Are the hardwood species presently on your land the best suited for your particular sites?
- What are the primary hardwood markets in your area?
- Have you identified overstocked stands that currently contain potentially valuable hardwoods?
- Do you have hardwood stands that have been degraded and need regenerating or intermediate treatments to rehabilitate them?

valuating Stands for Intermediate Treatments

There are a number of factors that affect when and how intermediate treatments should be dministered. The following factors should be evaluated or addressed prior to making silvicultural nd management decisions.

ite Productivity

To properly manage hardwoods, keen attention must be paid to properly matching species and soils. The success or failure of a forest management plan depends, particularly where timber is concerned, n correct evaluation of hardwood site productivity to ensure that intermediate treatments are conomically justified. Typically, hardwood sites can be grouped into three potential productivity classes: good, medium, and poor. The potential productivity of the site will dictate to a large extent species composition, ease or difficulty of getting desired regeneration, and the cultural practices necessary to develop high quality trees.

Good quality sites will typically grow oaks at least 75 feet tall in 50 years (oak site index = 75), and yellow-poplar 80 feet (yellow-poplar site index = 80). These sites are typified by deep (60 inches or more), moist, medium-textured (loamy) soils with a well-developed layer of topsoil more than six inches deep. These soils are adequately drained, with moisture-retaining subsoil. These soils occur most often along stream and river bottoms and upland coves and lower slopes and benches, especially on north and east exposures. These sites have the best potential for value growth.

Medium quality sites, typically growing oak trees 65 to 75 feet tall in 50 years (oak site index 65 to 75 and yellow-poplar site index 60 to 80), can also provide reasonable returns when intermediate treatments are prescribed. Sites producing less than 65 feet of oak growth in 50 years should be evaluated carefully before intensive intermediate treatments are prescribed. Sites with poor potential productivity are often found on mid and upper south and west facing slopes, ridges on upland sites, and areas with poor drainage.

While individual areas of homogenous soils can be large in bottomlands and in relatively flat coastal zones and areas like the Piedmont, site quality is often more variable in the mountains. Here, topographic features such as elevation, aspect, steepness, slope position, and slope length and shape strongly influence site quality. A list of good to medium hardwood sites and soil characteristics would include:

- Well-drained stream and river bottoms
- Mountain coves
- Benches (natural terrace formations on side slopes)
- Mid and lower slopes facing north or east
- Lower slopes facing northwest or southeast
- Gradual/concave slopes
- Soil more than three feet deep, with a well-developed layer of topsoil
- Medium-textured loams (little or no heavy clays or deep sands)
- Moist, well-drained topsoil with moisture-holding capacity
- Soils having less than 65 percent rock content

Stand Delineation

A stand is defined as an area that encompasses trees of a similar size, species, and quality or potential quality that are growing on soils with uniform productivity. The stand is the unit at which most silvicultural prescriptions are made. Proper delineation and description is important if intermediate treatments for high quality hardwoods are to be effectively applied. Forest maps should include stand boundaries, and plans should include stand descriptions. These descriptions should contain dominant species in the over and understories, size class and age of the majority of the growing stock otal basal area per acre, indications of the quality of the stems and of any problems that will impede and development.

latural Succession and Stand Development

rom a stand development point of view, intermediate treatments can begin as soon as trees in a egenerating age class have closed the canopy. Typically this is 10 to 15 years after a regeneration arvest, shorter on a highly productive site and longer on a site of poor productivity. Generally an npenetrable tangle of both desired and weed tree species, vines, briars and annuals occupy egenerating forests for the first five years. During the next five years a dramatic reduction of briars nd annuals occurs and shade-intolerant and intermediate species begin to become dominant.

Dominant species vary by landform. It is important to understand what forest types are best suited nd typically should develop on a given site. Table 11.1 shows dominant species for typical landforms n the South when an even-aged regeneration harvest is applied.

For coastal plain river bottoms, these species include eastern cottonwood, river birch, sycamore, ellow-poplar, sweetgum, green ash and cherrybark oak along with the more tolerant red maple. ligh quality upland sites such as coves, gulfs and north and east facing lower slopes, typically consists f yellow-poplar, cucumber magnolia, birch, white ash and northern red oak with tolerants including uckeye, basswood and red and sugar maple. In similar topographic positions above 3,500 feet, black herry and yellow birch combine with the tolerant sugar maple and American beech to become najor components of stands. In contrast, as you progress from these high quality sites to poor quality tes, oaks such as white, chestnut, post, and black as well as upland hickories and red maple become ncreasingly dominant. Additional information on hardwood site types can be found in Chapter even.

Regardless of the site, when crowns of dominant species grow together to form a single canopy, ntermediate treatments can be used to adjust species composition, select and culture crop trees, and emove unwanted and non-commercial species.

Jndesirable Stand Development

here are cases when a heavy harvest does not provide for adequate development of enough esirable stems to successfully manage. Examples of these include the following:

- Old-growth timber on swamp sites where sprouting rarely occurs, succession is often to bulrushes, cattails, black willow and ash, irreversibly destroying commercial productivity of the site.
- Elevated water tables in peat swamps and wet flats can cause reversion to gallberry, wax myrtle, sweetbay and red maple. Natural succession from this vegetation type to a more valuable hardwood timber type is slow and difficult to attain.

- High-grading on high quality sites, (red river bottoms, black river bottoms, branch bottoms, bottomlands and coves, gulfs and lower slopes) has caused shade-tolerant, commercially undesirable species such as ironwood, boxelder, and elm as well as suspect species such as beech and red and sugar maple to proliferate. Subsequent clearcutting of these degraded stands encourages sprout growth reproduction of the undesirable species in quantities greater than originally present.

In cases where undesirable development is occurring, an intermediate treatment may not be able to salvage the situation. This is especially so if noncommercial species dominate. In these instances regeneration and site preparation may be required.

Stand Structure and Age

Undisturbed hardwood stands usually occur as a single age class, with the main canopy trees varying by only 10 to 15 years in age. Trees within that single age class may be found in all crown classes (dominant, codominant, intermediate, and overtopped). In these instances it is extremely common for large diameter dominant trees to be the same age as the smaller diameter overtopped trees. It is important to remember that in many hardwood stands *diameter does not indicate age.* Multiple age classes (such as 20, 50 and 80 years) can occur as a result of repeated partial harvests. The major difference between single age and multiple-aged stands is that separation into crown classes is happening in each of the age classes, so that there may be young, large, dominant trees or small, old overtopped trees.

Stand structure has important management implications. Smaller trees, incorrectly thought to be young, may mistakenly be considered as trees to be retained until the end of the rotation. However, when small trees are old and have low vigor, they are unable to respond to thinning. Frequently, they further deteriorate under the stress of being released. It is therefore important to understand the age structure of your hardwood stands before prescribing treatments.

Tree Quality and Crop Trees

The issue of tree quality has been emphasized throughout the chapter. Development of a few high quality hardwoods per acre is often preferable to growing a large number of low quality stems. Because of this emphasis on quality it is important to be able to evaluate trees in a stand relative to their quality. Quality means that trees possess characteristics necessary to meet management objectives. For example, good characteristics for a wildlife objective might include the ability to generate mast and provide nesting habitat. However, the most stringent requirements for quality involve the production of high valued sawtimber or veneer. These trees must have main stems that are straight and free of branches or other indicators of wood defects, contain little or no rot, have a well formed and proportioned crown, and be a commercial species. Trees that possess these characteristics are termed "acceptable growing stocking" and for the production of timber are often

Drainage Class	Forest Site Type	Initial Species
Well drained	Cove, Gulf and Lower slope (High elevation)	Sugar maple, black cherry, yellow birch, American beech
	Cove and Lower slope (Low elevation)*	Buckeye, basswood, cucumber magnolia, yellow-poplar, black birch, red maple, white ash, northern red oak, black walnut
	Upper slopes, ridges, dry aspects (south, west)	white, post, chestnut oaks, scarlet oak, shrub oaks, hickories, black gum, red maple
	Bottomland*	Yellow-poplar, sweetgum, green ash, pump kin ash, cherrybark oak, pecan
	Red River Bottom	Cottonwood, willow, river birch, sycamore, cherrybark oak
	Black River Bottom and Branch Bottom*	Sweetgum, swamp blackgum, tupelo, yello-poplar, baldcypress, gallberry, switch can, red maple, wax myrtle, sweetbay, iron wood, water oak, willow oak
	Wet Flat	Switch cane, sweetgum, red maple, black gum, water oak, willow oak, cherrybark oak, ironwood, loblolly pine
	Peat Swamp	Switch cane, gallberry, wax myrtle, sweet bay, red maple, blackgum, pond pine, Atlantic white cedar, yellow-poplar
Very Poorly drained	Muck Swamp (Unimpeded drainage)	Tupelo, swamp blackgum, bald cypress, green ash, boxelder
	Muck Swamp (Impeded drainage)	Marsh grass, bulrush, cattail, black willow, swamp cottonwood, ash, boxelder

** Yellow-poplar and sweetgum occur commonly on berms of drainage ditches dissecting branch bottoms. Switch cane, red maple and water and willow oaks occur beyond the berm. Yellow-poplar is not a stand component in Mississippi river delta bottomland or Ozark highland coves.*

able 11.1 Species Composition Following a Biological Clearcut, by Soil Drainage Class and Site Type

selected as crop trees for management. Crop tree is a silvicultural term used for a tree that will be managed for and retained to the end of a rotation. To evaluate the crop tree characteristics for an individual tree, it is useful to classify the trees into size classes.

Sapling

Trees measuring 2 to 5 inches DBH are considered small trees or saplings. Typically these trees are not of commercial size, but if they are a desirable species and have good stem form and a well proportioned crown, they can be classified as a crop tree for timber objectives. Good stem form means that the main stem is straight (remember, this is your butt log) and unforked. Larger saplings should have branches that will be shed (or self-pruned) from the developing butt log. A well proportioned crown means that the tree possesses apical dominance (the crown has a distinct growing tip), 50 percent of the total height of the tree is in crown (50 percent live crown ratio) and the crown has developed on all four sides. If shade tolerant species such as beech, and sugar and red maple are being managed, these trees could be in the understory. However, typically shade intolerant and intermediate trees should be dominant or co-dominant in a regenerating area. They may be suitable to leave for 20 to 60 years for future intermediate cuts.

Pole

Trees measuring 6 to 10 inches in DBH are considered pole sized. As with saplings these trees must be the proper species and have a well formed main stem that is straight and solid with no indicators of rot. Typically all branches on the butt log section (lowest 16 feet), especially the lowest 12 feet, have died and only knots or overgrowths from branches are found at the top of the butt log section. These trees must still possess apical dominance and have full crowns. They can be retained for future intermediate cuts within the next 20 to 40 years.

Sawtimber

Trees measuring 10 inches or more in DBH are considered sawtimber. However, in some areas local merchantable limits will not provide for a sawtimber harvest until trees are 14 to 16 inches DBH. Desirable species, having good form and quality, may not be large enough to be considered mature, but will be satisfactory as crop trees in a final stand. Trees of this size should have butt logs that are well developed with a lack of defect indicators such as branch overgrowths on their stems. As trees are on the way to maturity some hardwood species like oaks lose their apical dominance and their crowns begin to round out at the top. Once these trees reach 80 to 100 feet tall loss of apical dominance is usually not a problem as they have reached their height growth potential. It is more important that these trees have crowns that have expanded on all four sides. With trees this size they will be carried to the final rotation age or possibly taken during a final intermediate cut within 20 years.

The evaluation of quality and designation of crop trees is an important part of developing many ypes of intermediate treatments. Essentially, you are designating the trees that you are spending noney to cultivate, and the payoff, monetary or otherwise, needs to be assured.

Many potential crop trees will be cut during intermediate treatments. Others will undoubtedly uccumb to insects, disease or other factors. *The guiding principle should be to favor and carry to final arvest the best 40 to 50 trees per acre.* This number is about the highest number of 20 to 24 inch ardwoods that a typical site can support. In some cases this number can be higher for species such s yellow-poplar that retain a conical shaped crown for long periods of time. All intermediate reatments and cultural work should be done to promote the growth and value of these better trees.

tocking

tocking refers to the relative density of trees per acre. Stocking guidelines are species and forest type pecific, and guidelines for different forest types have been developed to aid in maintaining the orrect density in stands to maximize growth. One hundred percent stocking means that the stand ontains the maximum number of trees of a given diameter for reasonable diameter growth.)ensities above this level are detrimental to long-term growth of trees in the stand. High stocking lso means that while individual tree diameter growth is not stagnant, trees are not growing in iameter rapidly. Typically stocking levels should be maintained at a level that allows trees to fully over the site, but also allows each tree in the stand enough room to grow a large crown, thus naintaining rapid diameter growth. Generally the minimum stocking level is 60 percent. If stocking s too high then thinning can be used to reduce the stand density to the appropriate level.

However, the question often is not: What is the total stocking of a stand, but, rather what is the tand stocked with? Is the stand stocked with high value trees or low quality trees or non-commercial pecies? Unfortunately, stocking is often a problem in hardwood stands. More often than not, these re remnant stands, following years of high-grading or diameter limit cutting, wildfires, and grazing. tocking in these stands is often inadequate and does not reflect the true potential of the site for ither quantity or quality. If the stand is severely understocked or stocked with noncommercial trees he best management alternative may be to harvest what remains of the stand and regenerate it. Iowever, if the stand contains adequate growing stock—trees of favorable species, age, and quality otential—then intermediate treatments can be used to enhance growth.

Stocking guidelines for different forest types have been developed to aid in this decision making. ypically foresters use these stocking guidelines to adjust density in a stand. Table 11.2 provides a implified version of stocking guidelines for hardwoods. This table indicates the approximate number f trees per acre by DBH class needed to carry adequately stocked stands through to your desired otation age.

This is merely a guide and does not mean that all the trees must be outstanding in quality and orm. For example, if a stand averages 6 inches DBH then 200 to 340 trees should be growing on an cre. However, as the stand develops to a 20 inch DBH, only 30 to 50 trees are needed to fully stock

DBH Class (Inches)	Stocking level Trees Per Acre	
	Low	High
6	200	340
8	140	240
10	90	150
12	70	115
14	50	100
16	40	90
18	35	60
20	30	50

Table 11.2 Stocking Table

the stand. Typically, not all of the 6 inch trees would possess the qualities to be a crop tree (a tree that you would manage and grow until mature). Let's say that only 90 of the 6 inch trees are acceptable growing stock. Is this good or bad? If you were interested in reaching a merchantable size of 20 inches note that you would need only 30 to 50 trees. The 60 you have would easily meet this demand, and the stand would be worth managing. With the assistance of a forester you can develop a crop tree release program to ensure that the 30 to 50 trees needed at the time of harvest would develop in the shortest possible time.

Evaluation Summary

The factors listed above must be addressed to ensure the effectiveness of intermediate treatments and to ensure that management funds are being applied in a manner that will provide the desired outcomes. As you read this chapter consider the following questions to help you identify the management needs of your woodlands:

- Does your management plan adequately indicate what sites (or soils) are capable of good to medium levels of productivity? These are the areas where the money spent on intermediate treatments will provide the best economic return to you.
- Have you adequately located and defined hardwood stands on your property? The stand is the basic unit for silvicultural decisions.
- Knowledge of the stocking level and composition of each stand is required to prescribe many intermediate treatments. In your stand descriptions, have you noted:
  - the overall stocking level
  - the size and age of trees occupying the majority of the stocking (main stand)
  - problem areas that have been degraded and need rehabilitation?
- Have you described the characteristics for crop trees in each stand? Many intermediate treatments in hardwood stands are geared towards cultivating a limited number of trees, some of which will ultimately be grown to the end of the rotation. Knowing how to select these trees is critical to effective use of intermediate treatments such as thinning and crop tree release.

Intermediate Treatments for Hardwood Management

Intermediate treatments are by definition any prescription that is developed to culture trees between regeneration treatments or final harvests. In naturally regenerating stands produced from a harvest these intermediate treatments typically can start as soon as stand development has produced a canopy. Intermediate treatments can then be used throughout the rotation to ensure proper species composition and good growth of the trees that will be carried through to the end of the rotation. Not only do intermediate treatments guide the development of the final crop trees, but intermediate treatments also ensure that effective use is made of the many trees that will not be in the final stand. Intermediate treatments can also be used in forests that have been subjected to light harvesting to improve development of both even aged and uneven aged stands.

There are a large number of intermediate treatments that can be classified in a number of different ways. However, most intermediate treatments can be grouped into one of the following:

- Cleaning or liberation—the removal of unwanted trees that are overtopping desirable ones
- Release—most commonly *crop tree release,* where individual crop trees are defined and each individual tree is released to provide increased growing space through the removal of surrounding trees
- Thinning—reduction in the density of well stocked stands to ensure adequate diameter growth
- Improvement and sanitation—the removal of unwanted trees, shrubs or vines that hinder the long-term development of stands
- Mid-story removal—the removal of understory and mid-story (overtopped and intermediate crown class) trees that are suppressing the development of advanced regeneration (typically oak)

Each type of treatment could be completed using a commercial harvest if enough merchantable material is being removed. If the trees to be removed are too small for local markets (pre-commercial) or when the quality of removals is unacceptable for local markets, the treatment is done without the aid of a harvest. In areas where pulpwood markets exist, it is much easier to complete intermediate treatments compared to areas where only sawtimber markets exist. In the latter instance, money often must be spent to implement intermediate treatments. Most forest owners try to develop intermediate treatments that can associated with a harvest, otherwise owners typically offset costs with funding from cost-share or assistance programs. When harvests are used, it is important to remember that the objective of the harvest is not to maximize revenue but to maximize the effectiveness of the intermediate treatment. It is also important to realize that several intermediate treatments can actually be combined in a stand. The following provides more information on intermediate treatments and examples of when and how they are typically used.

Cleaning or Liberation

This treatment is usually prescribed for regenerating stands where desirable species are overtopped by undesirable trees. One common use of this practice is after a commercial clearcut where no site preparation treatment was used during or directly after the harvest. In some cases an abundance of crooked, forked, or otherwise undesirable trees are left standing. These trees are above the young regenerating age class, and if enough are present (typically above 20 to 30 square feet of basal area per acre) their removal would be required to provide enough long-term growing space to the regenerating age class. Another situation where cleanings are used is when a newly regenerated forest becomes dominated by species of low commercial value. The overtopping trees are the same age as the desirables but have outgrown them.

Release

Release is used in stands containing potentially valuable crop trees that have inadequate growing space due to competing trees that are approximately the same height as the crop trees. The goal of this treatment is to ensure that crop trees have an adequate amount of growing space for crown expansion. Full crowns are necessary to provide for maximized diameter growth. Results of crop tree release research have shown that it is important to ensure that individual crop trees have 3 or 4 sides of their crown released to ensure long-term diameter growth. Figure 11.3 shows the release of a crop tree on three sides. The treatment is implemented by first identifying crop trees and then focusing on the removal of only trees of similar height that are competing with the crop tree.

Figure 11.4 shows how this treatment looks on a stand basis. Typically, a minimum of 50 crop trees should be left per acre in large sapling or pole sized stands. In small sawtimber sized stands 30 to 50 trees often can be found as crop trees. However, the total number of crop trees is often less important than making sure that release of the individual trees is done correctly. For example, a

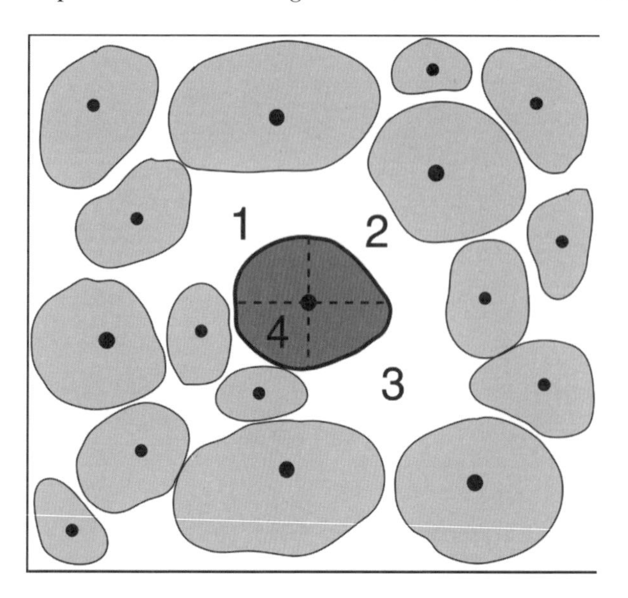

Figure 11.3 "Diagram of 3 sided crown touching release of a crop tree to maintain high levels of DBH growth."

Credit: Jeff Stringer

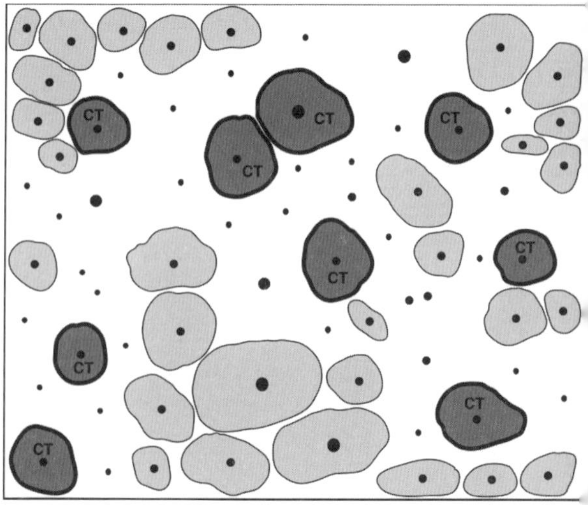

Figure 11.4 "Diagram showing distribution of crop trees released on 3 or 4 sides."

Credit: Jeff Stringer

ıixed stand of oak, yellow-poplar and maple is regenerating. The stand is now 20 years old and is ılly stocked. As is often the case, several of the yellow-poplars are taller than the other species, and ıeir crowns have achieved dominance and are free to grow on all sides. These trees, if desired, can e selected as crop trees and need no release because they are already free to grow. The oak and ıaple are competing heavily for growing space. Their crowns are touching, restricting lateral crown rowth of each individual tree. For a timber objective, the commercial oak species with good stem orm are selected as crop trees, and maples or poorly formed oaks that are touching the crowns of he crop trees are removed. This is termed a crown touching release and the end result is that the ellow-poplar and oak crop trees have at least 3 or 4 sides of their crowns released and are free to xpand their crowns on at least 3 sides.

A release treatment can be used for a number of different management objectives, not just timber. 'or example, if you are trying to keep a species dominant in a stand for wildlife purposes or because is rare, these trees can become crop trees and a crown touching release is used to ensure that they ave continued growing space. In these instances the number of crop trees per acre is not critical. Iowever, if timber is your objective you should not ignore stocking levels. Check release treatments ɔ make sure that adequate stocking is being maintained.

'hinning

'hinning may be appropriate where there are a large number of desirable trees that are too crowded or good growth. A release and a thinning are similar in that both reduce stand density. However, in he case of a thinning, there are a large number of acceptable trees present, and individual tree elections are not necessary. The goal is to reduce stand density, assuming that there are enough crop trees" present; their individual selection is not necessary. Sapling sized stands should be valuated carefully before a pre-commercial thinning is recommended. Generally, pre-commercial hinning is not necessary in young yellow-poplar or other overstocked stands of seedling origin. Many ıardwood species will express dominance without thinning because the physical site and neighboring rees determine site dominance.

As a general rule, for the first thinning, groups of hardwood trees should be old enough (at least 0 to 15 years) to allow time to ensure self-pruning of the first log (lower 17 feet of the tree) by ıaintaining close spacing of the trees. Use stocking guidelines to help determine removals. Generally, stand density should be regulated to keep stands between 60 and 80 percent stocked. Iowever, if stands are initially near or above full stocking, it is not always advisable to dramatically educe stand density to the desired final density in one operation. In these cases, make thinnings ight and frequent (at 5 to 10 year intervals). Remove only a few trees at a time to avoid epicormic prouting (sprouts that develop on tree trunks exposed to sunlight) on residual trees. Remove trees hat already have epicormic branches. Leave trees with large, vigorously growing crowns. After hinning, crowns of trees should be about 4 to 5 feet apart to provide space for expansion. There is ıo need to remove trees with crowns well below the level of crop tree crowns. These suppressed trees

Figure 11.5 "Trees with many epicormic branches should be removed when thinning your hardwood stands."

Credit—Dr. Michael Golden

serve as nurse or trainer trees, shading the trunks of crop trees to reduce epicormic sprouts.

Improvement

Many established hardwood stands have a high number of undesirable trees resulting from previous management practices. These stands can be rehabilitated by timber stand improvement (TSI) which includes cutting out undesirable or competing tree species, either in the understory, overstory, or both, and controlling other unwanted vegetation. The purpose is to give desirable trees a chance to grow better. Ideally TSI should be accomplished in stands on good to medium quality sites with at least 30 to 50 good crop trees per acre.

A harvest can sometimes be used to implement an improvement treatment to salvage trees that are undesirable species, have poor form, are diseased or insect-ridden, or are slow-growing or overmature. Epicormic branches along the trunk, fire scars and butt rot (rot in the lowest 16 feet of the bole) reduces timber quality. Wildfire or logging damage during partial harvest can cause scars in the bark or wood of the remaining trees. The injured stems eventually may develop rot. When using an improvement cut, be sure to emphasize the importance of reducing logging damage to residual trees. Trees may be sold or used by the landowner for firewood, fenceposts or other purposes. In marking trees for TSI, the main rule is "Save the Best." To upgrade a stand, first cut out dead, diseased and crooked trees. Next, remove trees with flat-topped crowns because they have poor growth potential. A tall, straight trunk with a full, rounded crown of branches indicates healthy growth. Deeply fissured bark in some species indicates good growth in older trees.

Some stands need to have problems such as grape vines removed when they are in high enough concentration to interfere with future regeneration. Other vines, shrubs or trees that are a nuisance and exotic invasive species need elimination. This work is often accomplished as part of a TSI treatment.

d-Story Removal

any forest owners are interested in maintaining oaks in their forests for both wildlife and timber als. To ensure future oak regeneration it is important on most good to medium quality sites to ve oak advance regeneration present at the time of a regeneration harvest. Advance oak generation are large seedlings (over 4 feet tall) or small saplings present prior to a regeneration t. These must be present for oak to have a head start and compete with fast growing competing ecies like yellow-poplar or sweetgum. Unfortunately, the development of oak advance regeneration ust be initiated several years before a regeneration harvest, in most instances 5 to 10 years. It has o been shown that on most good and medium quality sites adequate oak advance regeneration ust be cultivated with the aid of an intermediate treatment. The main factor that restricts oak vanced reproduction on these sites is shade from understory and mid-story trees (typically maple, ech, or other shade tolerant species). Oak stands will sporadically generate small oak seedlings. owever, these seedlings growing in deep shade will stay small, quickly lose vigor and eventually die. removing understory and mid-story trees enough filtered light is available for many species of oaks develop into large seedlings or saplings. After the development of this vigorous advance oak production, the regeneration harvest can be scheduled. The treatment should be used on good to edium quality sites that have small underdeveloped oak seedlings present.

ethods for Applying Intermediate Treatments

nce the intermediate treatment has been decided upon it is necessary to tackle the issue of how to t rid of unwanted trees. There are numerous mechanical and chemical methods available.

echanical Methods

oviously, if a commercial harvest can be accomplished the harvest may pay for the treatment. void damaging healthy residual trees while felling and removing poor trees.) However, it may be tter to leave a large tree standing and kill it by girdling or using chemicals rather than risk maging potentially valuable adjacent trees. Also, many forms of wildlife will use large dead or dying es for dens, nests, and perches. There are also other situations when the intermediate treatment nnot be completed using a commercial harvest. In these cases the trees must be cut or they must deadened either mechanically or with the use of chemicals (see below). Mechanical methods quire only the use of an ax, brush saw, or chainsaw. Small trees can usually be cut down. However, extremely dense stands where a large number of small trees need to be killed or where sawtimber ed trees are to be eliminated, girdling is often a better choice. Girdling is one of the oldest chniques used to get rid of unwanted trees. This method eliminates the undesirable tree, but avoids maging the young understory trees because the dead girdled tree disintegrates slowly. Typically, rdles are cuts one-half to one inch into the wood (under the bark) and continuous around the tree. hen using girdles to kill vigorously growing sapling and pole sized trees a double girdle should be ed. This is because vigorously growing trees can successfully callous over a single girdle. Oaks can be killed with a single one-inch deep girdle regardless of their size and vigor because only the outer few growth rings in oak conducts the majority of the water to the crown, and severing the water supply will typically kill the tree before it has a chance to callus over a girdle. Girdles typically do nc kill the root system of younger trees and they will resprout following girdling. This often is not a problem unless you need to eliminate shade tolerant species. If you want to remove these species or use a more cost effective method of removing unwanted trees there are several chemical applicatio methods that effectively control individual trees and understory vegetation.

Chemical Methods

Herbicides can be used to control or kill the entire tree including the root system, and treatments using chemicals are often faster than mechanical treatments. Herbicides may be applied to individu trees by injection, basal application, or stump spraying. A successful herbicide application requires the selection of the best chemical, using the correct application technique, and applying at the proper rate and at the best time. Flashback, as was mentioned in Chapter Ten, is also a consideratio in herbicide selection.

Selection of the best herbicide will depend upon the species to be controlled, and the cost. Tree species vary in sensitivity to herbicides, and herbicides vary in cost. You must match target species tolerance with chemical costs.

There are several methods for applying herbicides to individual stems, and selecting the proper technique depends upon many factors. Tree injection is preferable for trees of two inches in diame or larger. Injection equipment can cost from $200 to $300 and is available from any forestry supply company. A hatchet and a squirt bottle provide an effective low-cost alternative to tree injectors. Application rates for injection are provided on the herbicide label, which should be carefully reviewed prior to use. Generally, one ml. of the herbicide (or a diluted herbicide solution) is inject into evenly spaced cuts made through the bark and into the cambium layer. About one cut (one ml is needed for every two to four inches of tree diameter. Injection should *not* be used in early spring during the period of maximum upward sap flow.

Trees that are smaller than two inches in diameter can be controlled by directed foliar sprays. Spr herbicides on the foliage of individual crowns to the point of runoff, being careful not to spray desirable adjacent trees. It is easy to apply too much chemical. Typically the leaves of treated plants should look as if they have a mist of the herbicide solution on them. Follow label directions regarding mixing percentages and any regulations pertaining to the use of herbicides.

A second alternative for individual treatment of smaller stems is a basal bark application using a stream-line technique or a full basal spray. A backpack sprayer is used to apply the herbicides to the tree stem at between 6 to 18 inches above the ground. Although more easily directed to a single tre application costs per acre may be high when there are many undesirable trees to treat.

When trees are cut sprouts can be controlled by a stump treatment. The herbicide should be sprayed on the outside two to three inches of the stump surface and topmost two to three inches of

: bark as soon after cutting the tree as possible, typically within one to two hours. The stump
gins to dry out and forms callus barriers immediately after cutting, and a delay of a few hours in
atment can result in a significant loss of control. If older stumps need to be treated chemicals used
basal bark application can be used. However, it is best to use the appropriate chemicals on a
shly cut stump.
Sensitivity of weed and crop species to the herbicide should be indicated on the container label.
th timing and application rate are critical in chemical release operations. Improper application
y result in damage to neighboring farms or crops, damage to your trees, ineffective weed control
a health hazard to others as well as the applicator. A list of chemicals that have been used for
desirable species control is included in Table 11.3.
Any chemical you select for vegetation control must be labeled for forestry use in your state. Read
available literature, especially the container label before use. The label includes information about
ety, application rate, methods and conditions of application and the effectiveness of the chemical
approved uses. In addition, you may need a herbicide and/or a pesticide applicator's license. This
ormation can be obtained from your county agent, and your county agent can answer questions
d provide other information regarding your particular situation. Contact a professional forester
d consider contracting herbicide application to a certified company.

plication Costs

e cost of chemical application is based on the number of stems per acre treated and the size and
ecies of the trees treated. These factors affect the type of chemical used, how much chemical is
ed, and how quickly it can be administered. While chemicals vary significantly in cost per gallon
d the amount of chemical applied per acre, the application time is a significant factor that should
o be taken into consideration when determining the total cost of chemical treatments. For
ample, a review of labels and costs for chemicals typically recommended for tree injection indicates
at the chemical cost per square foot of basal area treated is approximately the same (approximately
.00). However, application times are significantly different. Table 11.4 indicates costs per square
ot of basal area treated (using the highest label application rate) and the relative time it takes to
ply the chemicals based on label guidelines.

re

escribed fire has limited utility for hardwood management. While there is significant interest in the
e of prescribed fire for ecological purposes in hardwood forests, there are significant operational
rdles that must be overcome if fire is used for applying intermediate treatments in hardwood
nds where sawtimber or veneer is a goal. Fire is generally not used in established hardwood stands
cause it can easily damage the relatively thin bark of hardwood trees. Not only can fire kill smaller
rdwood stems, it may damage larger stems and create openings for pest and decay organisms. If
e is being considered as a tool to implement an intermediate treatment, fuel load and fuel

Product	Label Use		
	Inject	Basal	Stump
Access	No	No	Yes
Accord	Yes	No	Yes
Arsenal	Yes	No	Yes
Banvel	Yes	No	Yes
Banvel CST	Yes	No	Yes
Chopper	No	Yes	Yes
Garlon 3A	Yes	No	Yes
Garlon 4	No	Yes	No
Pathfinder	No	Yes	Yes
Pathway	Yes	No	Yes
Remedy RTU	No	Yes	Yes
Rodeo	Yes	No	Yes
Roundup	Yes	No	No
Super Brush Killer	Yes	Yes	Yes
Tordon 101M	Yes	No	Yes
Tordon 101R & RTU	Yes	No	Yes
Weedone 170	No	Yes	Yes
Weedone CB	No	Yes	Yes

[1]This list may not be complete, nor does the inclusion of an herbicide constitute a use recommendation or imply labeling for forestry usage. Some herbicides are prohibited in certain states, or are labeled only for minor uses, or are labeled only in certain states under state local need labels (FIFRA Sec. 24C). Some herbicides or situations may require licensing or certification for purchase or application. Contact your local Extension agent, county forester or forestry consultant for specific recommendations in the state and area where the herbicide is to be used. This listing should be considered a guide for preliminary planning.

Table 11.3 Herbicides Labeled for Tree Injection, Basal Application and Stump Sprays in the South[1]

Product	Rate[1]	(Per Sq. Ft. Basal Area)		Relative Application Time[3]
		Gallons	Dollars[2]	
Garlon Remedy	0.5 ml at 3–4 inch intervals	0.0163	1.195	57%
Accord Roundup Glyphosate	1 ml per 2 inch DBH	0.0184	1.091	64%
Pathway Tordon	1 ml per slit 2-3 inch apart	0.0285	0.913	100%
Arsenal Chopper	1 ml per 3 inch DBH	0.0024	1.178	15%

[1] Assumes highest label rate.

[2] 2004 dollars.

[3] Percentage is based on the slowest application time. Applying Pathway and Tordon takes the longest of the chemicals reviewed and their relative application time is 100 percent. All other chemicals are a percentage of the time taken for Pathway and Tordon.

ble 11.4 Rates and Times for the Application of Common Tree Injection Chemicals

oximity to crop trees or other valuable trees must be considered. Prescribed fire may be used as a e preparation tool where numerous noncommercial stems remain after harvest. A fire encourages routing from below the ground, which places all species on a more or less equal footing when mpeting for site resources.

atural Stand Protection

ecause of the long rotation ages and the premium market prices associated with high quality rdwoods, the most critical management opportunities generally deal with stand protection. undamental concerns include control of grazing, wild hogs, and beavers, logging damage and otection from fire and soil damage (compaction, blocked drainage and erosion). Maintenance of gor and species diversity may also offer some protection from insect and disease damage. Please view Chapter Nine for additional information on forest protection.

lantation Management

any landowners who have seen the attractive prices paid for quality rare hardwoods have nsidered establishing plantations of those species. This category of trees usually starts with black walnut and includes cherry, some oaks and occasionally ash or pecan. Exotic species, such as Paulownia and hybrid poplars, attract similar attention. Table 11.5, below, gives the values of final products from a single hardwood tree.

Frequently, high market prices have resulted from scarcity and traditional preferences more than from any particular special utility. Reasons for scarcity may include extremely demanding site requirements, environmental problems or history of exploitation. As a result, market trends may be very unpredictable as tastes change or substitution occurs. Hardwood plantation management shoul be undertaken with caution and commitment. The plantation site must be carefully selected and matched to a suitable species. The site must be intensively prepared to control weeds before plantin The same extremely demanding site requirements that have contributed to the scarcity of walnut, fc example, lead to luxuriant growth of competing weeds and vines. That competition must be controlled for several years to get acceptable establishment, growth and stem form in crop trees.

Straight stems without branches are necessary to meet most hardwood market quality standards, s pruning may become a required management practice in widely spaced plantations. Because of the labor intensity, pruning should be restricted to the first log of crop trees only.

Crop tree spacing, inter-planting of species, potential rotation ages and specific cultural recommendations all depend upon the site, species and product produced. Publications describe intensive management techniques for the most common situation, but these must be customized to the individual site and product, as well as to the growth, quality and problems confronted in each plantation.

Size[1]				Units	Value of Products ($)[2]			
Tree Age	DBH (in.)	Merch. Height	Cords	Board Feet	Fire Wood	Pulp Wood	Saw Timber	Prime Saw Timber
10	5	10	0.02	—	1.00	0.66	0.00	0.00
30	10	15	0.04	18	2.00	1.32	0.00	0.00
50	15	24	0.24	124	12.00	7.92	37.20	0.00
70	20	32	0.49	296	24.50	16.17	118.40	0.00
90	25	40	0.98	574	47.50	32.34	229.60	459.20
110	30	48	1.53	991	76.50	50.49	396.40	792.80

[1]*Assumes tree grows 1/4-inch each year in diameter.*

[2]*Firewood prices based on $50/cord. Pulpwood prices estimated at $33/cord. Sawtimber prices estimated at $400/MBF. Prime sawtimber prices estimated at $800/MBF. (MBF = Thousand board feet)*

Table 11.5 Values of Final Products Produced from a Red Oak Tree Based on Tree Age, Tree Size and Deliver Product Price

mmary

nagement of existing stands generally includes only those activities that protect, upgrade the ality, or improve the growth rate of residual trees, especially for timber production. Crop trees ıst be identified by their anticipated performance and merchantability, which depends on species, or, form, and quality. Stands with inadequate numbers of crop trees should be regenerated at the liest opportunity. Cleaning treatments are used to release crop trees from overtopping vegetation. lease treatments are used to provide growing space for crop trees on an individual tree basis. erstocked stands may respond to thinning, and when harvesting is used it must be performed efully, avoiding damage to residual trees. Those stands, which are adequately stocked but contain ny defective trees and other problems (such as the occurrence of invasive exotics) may benefit m timber stand improvement. And finally, future regeneration of oak may be enhanced through use of a mid-story removal treatment. Improved markets for firewood and other low quality and all material will provide additional economic opportunities for natural hardwood management. It mportant to remember that proper hardwood management, especially for timber production, ıst couple the use of intermediate treatments with proper regeneration and site preparation atments to adequately shape future stand structure and growth. Deliberate management of high ue hardwoods in single-species plantations continues to be speculative because of uncertain rkets, long rotations and site sensitivity and variation.

view Questions

1. Why does delineating stands and defining site productivity have important management implications?

2. What is stocking and what number of crop trees is typically required to meet full stocking levels at the end of a sawtimber rotation?

3. What are the important characteristics for sapling and polesize crop trees where timber is the primary objective?

4. Describe a crown touching release.

5. What is the difference between thinning, release, and improvement treatments?

6. What risks are involved in hardwood plantation management?

7. What are the advantageous and disadvantages of using chemicals to deaden trees in intermedia treatments?

Suggested Resources

Readings

- USDA Forest Service, Southern Region. 1994. "Southern Hardwood Management." *Management Bulletin R8-MB 67.* Atlanta, GA.

Unit V: Marketing Timber

Most timberland owners will have the opportunity to harvest timber from their lands at some point during their ownership. Unfortunately, forest landowners are often at a disadvantage when they attempt to market their timber products, primarily because of their lack of experience in selling timber. Often, they are not familiar with the current market conditions, methods of marketing, legal and tax considerations, environmental considerations, timber buyers, and harvesting firms in their area. Knowledge of each of these aspects is necessary to maximize your immediate and long-term financial returns. The following chapter provides you with an overview of each of these subjects and guidance on how to obtain additional information for particular activities.

The initial step in any timber sale is for you to decide what objectives you want to meet:

- What do you want to accomplish with this sale?
- Is it a final harvest of your crop?
- Is it to improve the condition of your stand by removing the undesirable trees?
- What do you want to have when the harvest is complete?

These questions are the heart of the planning process. A plan for your timber sale is not only for the immediate future, but it must also consider how the planned harvest fits into the overall plan for your property as well as your other investments and income. This requires that you set your objectives and plan for your desired income requirements, tax situation, investment and estate plans, and long-term ownership goals.

The buying and selling of timber is usually an individual, local transaction as compared to other commodities that are generally priced on a national or even international basis. Most commodities are sold at a delivery point whereas timber is usually sold as it stands on your land. As a result, variations in timber quality, potential product yields, harvesting costs, and transportation will all influence the value of your crop to potential buyers.

Timber owners can be sure that buyers have a good idea of the values and costs involved on any particular timber tract. The first step you must take toward effective timber marketing is to seek advice from a professional forester and possibly your accountant and attorney. A forestry consultant can provide you with a variety of *marketing services,* including appraisals, timber management, timber marketing and timber sales administration. It is almost universally true that a professional forestry consultant can get much more for your timber than you can by selling it yourself. Their fees are usually minimal compared to the extra income they gain for you. See the suggested readings at the end of each chapter and the appendices at the end of the book for more information regarding how and where to obtain technical assistance for marketing your timber.

Chapter 12:

Marketing Your Timber

Once the overall objectives in managing and arketing timber are established, you can begin the process of offering timber for sale. Professional dvice in all aspects of marketing timber is available. The importance of utilizing professional ssistance can not be over-emphasized. This chapter identifies six steps in the marketing process and e issues that you need to consider in taking each step.

easuring Timber

ne of the confusing factors for many forest landowners is how standing timber is measured. ulpwood and chip-n-saw timber is often conveyed by weight, units of *cords,* or *"cunits."* A standard ord is a stack that measures 128 cubic feet (generally four feet high, four feet deep and eight feet ng). This includes bark, wood and voids and is an indication of space occupied rather than actual ood measured. The *cunit* refers to 100 cubic feet of solid wood rather than a stacked volume. (C is atin for 100, thus a c-unit or "cunit" equals 100 units.) During the past 30 years most pulp and aper mills have adopted weight scaling as a measure of volume. Each mill has its own conversion ctors for units of measure for the various species bought.

Saw and veneer logs are generally traded by *board-foot* units. A *log rule* estimates the board foot olume of either trees or logs. The log rules commonly used to estimate volume are *Doyle, Scribner,* and *International 1/4-inch.* The particular rule used varies by location within the state and the products being traded.

Each rule has a different estimated volume yield for each size log or tree because of the way the rules were formulated. For example, if the International 1/4-inch rule is used as the basis for the volume measurement, the Doyle scale in comparison will be up to 28 percent less, and the volume according to the Scribner scale will fall in between these two. Thus, a landowner must know what log rule is used if timber is sold by the unit. Comparing different per-unit offers for a stand of timber can be difficult if the offers are based on different log rules. The advice of a professional forester is necessary if the landowner is not well versed in the various log rules and their relationships to each other.

Scaling by weight is also commonly used for measuring yellow pine sawtimber. However, conversion from stick scaling to weight can be difficult when trying to calibrate the weight of X amount of logs that will produce Y amount of board feet of lumber. Log weight depends on many variables, but *specific gravity, moisture content, defect* and *bark* are generally the most important factors.

How Timber is Measured

The first step in estimating the volume of marketable timber on a given tract of land is to conduct a timber *cruise.* A preliminary inspection by someone from one of the public agencies can be a help in determining the marketability of the timber being considered, but an exact measure of the various timber products must usually be obtained by employing the services of a consulting forester. Most often, a systematic sample of trees representative of the entire stand is cruised. However, where stands of large or high-volume timber are present, all merchantable trees may be measured during a cruise.

The diameter breast high, or *DBH* (the diameter of the tree at a height of 4.5 feet above ground level), and the merchantable height in terms of 16-foot and additional 8-foot logs are generally the two measurements taken of each tree included in a timber cruise. The board-foot volume of a specific tree can be determined from these two measurements using assorted tree volume tables. The number of logs (usually 16 feet long each) to some merchantable top of the tree (usually a small-end diameter of 7 inches or more, inside the bark, or to a defect such as excessive knots or limbs) is the measured height of sawtimber. The tree height for pulpwood is normally recorded as the number of pulpwood sticks or feet to a merchantable top (usually a small-end diameter on the outside of the bark to 4 inches). Total tree heights may be counted instead for pulpwood.

If You Want to Try Measuring Yourself…

Trying your own timber cruise can provide first hand experience in calculating and estimating product volumes and can assist you in getting an overall feel for what exists in your stands. However, timber cruising is both a science and an art and one must participate in years of education and experience in order to be confident enough in one's own abilities to conduct an accurate timber

:ruise. For accurate timber cruises associated with timber sales or land purchases,
nire a professional forester.

A basic tool for measuring the boardfoot volume of standing trees is a calibrated
:ommonly referred to as a *Cruiser's stick* or *Biltmore stick* (scale stick). To measure th
squarely in front of the tree and hold the scale stick flat against the tree at a height 4.5 feet above
ground level and 25 inches from the eye in a position horizontal to the ground. Twenty-five inches
rom the eye roughly corresponds to the length of an adult's arm. Move the stick right or left until
he zero end of the stick lines up with the left edge of the tree trunk. Keeping your head still, read
he diameter figure that corresponds with the right edge of the tree trunk (DBH is an outside
neasurement). Diameters are usually measured to the nearest inch, but 2-inch diameter classes are
used where large numbers of trees are scaled. Measure the tree from the upper side when on sloping
ground. Because most trees are not perfectly round, taking 2 measurements at right angles to one
nother will provide a more accurate measurement.

Measuring the number of merchantable logs can also be done using a scale stick. Holding the stick
n a vertical position 25-inches from the eye and standing 66 feet away from the tree being measured,
he number of merchantable logs from the stick margin can be recorded. It is important to stand 66
eet from the tree being measured since the scale stick has been calibrated for that distance from the
ree. As with DBH, keep your head still when taking measurements. Tilt your head back slightly when
neasuring from stump point to the merchantable top so you do not have to reposition your head.

Scale sticks can be purchased from forestry equipment outlets such as Forestry Suppliers,
ncorporated (*www.forestry-suppliers.com*, phone 1-800-647-5368) or Ben Meadows (*www.benmeadows.com*,
hone 1-800-241-6401).

stimating Values

tanding timber is referred to as *"stumpage."* Current stumpage values and trends are available
hrough local extension offices, forestry agencies, and consulting foresters. The values are influenced
y many factors that change over time and are different for each tract. Of course, the tree species on
tract is a major influence of value. Each species differs in wood characteristics and thus its potential
nd use. Obviously, species in high demand and limited supply, such as high-quality white oak, will
ommand a higher stumpage price than a low grade species that can be used only as fuel.

The size and quality of the standing tree influences the value of the products that can be obtained
om it. Small, knotty, and crooked trees yield less product volume and products of lower value. Tree
ze also influences logging and milling costs because more pieces of small timber are handled per
nit of measure, increasing per-unit costs.

Logging cost and distance to utilizing mills have a great influence on the value of standing timber.
ransportation costs make up a significant percentage of the total harvesting costs faced by loggers.
he logging techniques used and the physical condition of the area can also affect the value
gnificantly. Adverse ground conditions such as wet swamps or steep terrain increase the cost of

sting. These aspects cause sales in wetter times of the year to command better prices for stands at are accessible and located on well-drained soils. A direct relationship also exists between the delivery distance and the cost of transportation. Road and bridge restrictions, brush conditions, and any other factors that increase logging costs will reduce the standing timber's value in the buyer's eyes.

An important factor in stumpage value is the competitive bidding by primary processing mills in the immediate buying area. Higher market prices may be obtained where several buyers are competing for a particular species and size of timber. This can be observed in regions where owners of pine sawtimber have a distinct market advantage because numerous mills exist. *Studies have shown that, in general, the more bidders there are bidding on a tract of timber, the higher the winning bid will be.*

The total acreage and volume of the sale, as well as the volume per tree, generally influence stumpage prices offered. Large tracts of high timber volume make it possible to reduce transportation costs per unit. This is associated with highly capitalized logging operations. Also, be aware that each different harvesting site has certain fixed costs associated with it. These include obtaining access, establishing trucking routes, and finding the best markets for all timber on the acreage.

A forest landowner will also find that the more stipulations and restrictions that are placed on the buyer and logger, the lower the price they can offer. Many of these restrictions increase the cost of harvesting. Thus, you must balance your contractual demands with the decrease in returns such demands could bring.

Product Specifications

Each tree must meet specifications or standards to qualify for conversion to a particular forest product. A general familiarity with these specifications will help you decide to what market your sale should be oriented. Usually your best price is realized and the highest degree of use achieved when trees are cut for the products for which they are best suited. This sounds obvious, but many times small-sized sawtimber is cut just for pulpwood. Knowledge of product specifications is helpful in planning and supervising the cutting operations.

Almost all wood products must be sawn from live, sound, reasonably straight trees free of burned o charred wood and free of nails, wire, or embedded metal. Crooks and forks may not be acceptable for some *sawn* products.

Over the years, different grading methods have tried to relate tree and log surface characteristics t product quality and yield. There are no universally accepted standards regarding tree and log grades Local practices usually dictate the grading method.

One such method for evaluating southern pine trees and logs was developed by the Southeastern Forest Experiment Station of the USDA Forest Service. Southern pine *trees* intended for yard and structural lumber using this scaling method should have a minimum DBH of 9.6 inches and a minimum merchantable height of two 16-foot logs. A southern pine tree would receive a grade of "A it had 3 or 4 clear faces on a graded section 16-feet in length. A *face* is defined as an area on the raded tree 1/4 the circumference of the graded section and extends the entire 16-foot length. A *ear face* is one free from knots measuring more than 1/2 inch in diameter, overgrown knots of any ze and holes over 1/4 inch in diameter. If the graded section has a *sweep* (the amount of curvature a tree or log as compared to a flat plane) in the lower 12 feet that is 3 inches or more and equals r exceeds 1/4 of the DBH, the tree is downgraded one letter grade from a grade of A or B. If lvanced heart rot is evident anywhere on the tree stem, the letter grade of A or B is also reduced by ne grade. Grade B trees are those with 1 or 2 clear faces on a graded section 16-feet in length, and rade C trees are those with no clear faces within the graded section.

Logs can be graded in a similar fashion. As an example, the Southeastern Forest Experiment Station f the USDA Forest Service recommends that southern pine logs intended for yard and structural mber be at least 6 inches in diameter inside the bark, have a net log scale of at least one-third their ross log scale, are 8 to 20 feet in length and contain only one 16-foot log if using this scaling ethod. Rather than grades of A, B or C, logs are assigned grades of 1, 2 or 3, respectively. The rading criteria for trees regarding clear faces, the amount of allowable sweep and heart rot are lentical for logs.

The USDA Forest Service's scaling method has not received widespread use, but some sawmills have lopted similar methods of evaluating quality. Hardwood species and other softwood species are also raded and have similar criteria.

pecifications List

he specification highlights included in this section cover only a few of the more common forest roducts. They are intended only to assist you in making a more informed judgment about the roduct potential of your timber. By no means are they sufficiently detailed to enable one not already horoughly familiar with timber-cutting practices to cut a tree into acceptable products. Furthermore, very mill has its own individual requirements, and no timber should ever be cut with the thought of elling it to a particular mill without first getting an order for the wood and a copy of that mill's wood pecifications.

hort Pulpwood

- *Species:* Pine and hardwood. Some buyers exclude certain species.
- *Diameter:* Minimum small end diameter varies by mill. Maximum varies by mill.
- *Length:* Usually 5'3", but varying from 5' to 10'. To be merchantable, a tree should contain at least 2 pulpwood bolts (lengths).

ongwood Pulpwood

- *Species:* Pine and hardwood. Some buyers exclude certain species.
- *Diameter:* Minimum top diameter varies by mill from 4" for cut long lengths to no minimum for tree length. Minimum diameter at 4-1/2' above the ground varies from 4" to 6".

Maximum 18" to 34" varying by mill.
- *Length:* Minimum 8' to 27' varying by mill. Maximum tree length must conform to highway regulations.

Sawlogs
- *Species:* Pine, hardwood, and other softwoods.
- *Diameter:* Minimum 6" to 8" inside bark at the small end for pine and 10" to 13" for hardwoo and cypress, varying by mill.
- *Length:* From 8' to tree length, varying by mill. Cut logs are measured in even-foot lengths plus a trim allowance. The most common lengths are 12'6", 14'6", and 16'6". To be merchantable, a tree should contain at least one 16' log.

Chip-n-Saw Logs
- *Species:* Pine.
- *Diameter:* Minimum 3"–6" inside bark at the small end, varying by mill. Maximum around 15" varying by mill.
- *Length:* Standard log lengths to tree length. Chip-n-saw trees are intermediate in size betwee pulpwood and sawtimber and contain varying amounts of each product, which are typically used for 2×4's, 2×6's and other building products.

Veneer Logs
- *Species:* Pine and hardwood.
- *Diameter:* Minimum 10" inside bark at the small end for pine. Minimum for hardwood is generally larger than 14" but varies by mill and method of manufacture.
- *Length:* Minimum 8' plus trim for hardwoods; 12' and 16' lengths plus trim are common. Pi veneer logs are normally 17' to 18' plus trim. Length requirements and trim allowances vary by mill and method of manufacture.

Poles
- *Species:* Pine.
- *Diameter:* Minimum top diameter varies from 3.8" to 8.6" inside bark, depending on length and class of pole. Minimum bottom diameter (measured 6' from the butt) varies from 6.4" t 17.2" inside bark, depending on length and class. Maximum bottom circumference cannot exceed specified minimum by more than 7".
- *Length:* Minimum 20'. Poles are usually cut to length in multiples of 5'. There are 10 classes poles determined by size and quality.

Piling
- *Species:* Pine.
- *Diameter:* Minimum top diameter varies from 5" to 9" inside bark at the small end. Minimum bottom diameter (measured 3' from the butt) varies from 12" to 14" depending on length and class. Maximum bottom diameter varies from 18" to 20", depending on length and class.

- *Length:* Minimum 30'. Piling is cut in multiples of 2' up to 40' in length and in multiples of 5' for lengths over 40'. There are several classes of pilings based on intended use, size, and quality.

'rossties

- *Species:* Oak, gum, pine. Many other hardwood species are also acceptable.
- *Diameter:* Minimum 10" inside bark at the small end for regular ties; 12" for switch ties.
- *Length:* Minimum 9'. Regular tie logs are cut in 9' or 18' lengths. Switch tie logs are cut in 12', 14', or 16' lengths plus trim. There are 4 classes of regular ties based on size and quality. Switch ties are not graded.

reparing the Site

ou and your forester should make arrangements for access to the tract, establish boundaries, and ecide on road locations and other "on-the-ground" preparations for a sale. Neighbors should be nade aware of the planned harvest, and adjoining property lines should be agreed on.

reparing a Contract or Sale Agreement

. timber sale often involves a substantial sum of money, and timber sales are something the typical ndowner does infrequently. Anyone considering a timber sale should not rely on personal idgment, but should confer with a forester and an attorney experienced in timber sales.

When selling timber, landowners should be certain that their interests are protected. Professional orestry assistance is well worth the cost when dealing with deeds, contracts, and other legal ormalities.

tanding or Severed Timber

ertain legal and tax implications determine whether timber should be sold standing or severed. *tanding timber in most states is real property, and cut timber is personal property.* By law, the conveyance of n interest in real property (standing timber) must be by written document.

Cutting contracts and deeds frequently provide the means of conveying rights to an interest in mber. Timber deeds are used most often when payment is made for standing timber, whereas ontracts are used when payment is for severed timber.

ontracts

. contract establishes the conditions to which buyer and seller agree as well as their rights and duties nder those conditions. Timber sales contracts or cutting contracts are commonly used in ansactions where timber is not sold in lump sum. Payment is made periodically as the timber is elivered to a mill or loading site, although title to the trees passes to the buyer once timber is evered.

Contracts are suitable for unit sales such as pulpwood, chip-n-saw, selective timber cuts, poles and pilings, and other specialty product sales. However, since standard contract forms seldom include provisions that a landowner may wish to include, a special document or stipulations may be needed. These provisions should state those items that buyer and seller have agreed on.

No two timber-cutting contracts are exactly alike, but all contracts should include the following basic provisions:

- Guarantee of title or right to sell, description of the sale, and location of boundary lines.
- Specific description of timber being conveyed, method of designating trees to cut, and when, where, and how to determine volume.
- Terms of payment.
- Duration and starting date of agreement.
- Clauses to cover damages to non-designated trees, the residual stand, fences, ditches, streams, roads, bridges, fields, and buildings.
- Clauses designating road location, construction requirements, maintenance, condition requirements following harvest, etc.
- Clauses to cover fire damage where harvesting crew is negligent and to protect seller from liability that may arise in the course of harvesting.
- Specifications for complete use of the merchantable portion of trees.
- Clauses to cover compliance with *Best Management Practices (BMPs)* such as protection of soils, establishment of water bars on roads, establishment of filter strips along streams, protection of drainage systems, etc.
- Penalties for littering or trash left on the tract.
- Clauses for arbitration in case of disagreement.

A good timber sale agreement is understandable, workable, and enforceable for both buyer and seller. However, it should not be so complicated that it attempts to cover all eventualities. Neither should it be so brief as to exclude essential points of the transaction.

Deeds

Deeds are used most often when all timber within a prescribed boundary is to be cut. Deeds are especially popular with purchasers of standing timber who usually require a deed upon payment. For example, most timber buyers prefer a timber deed prepared by their legal department. In many ways deeds are usually less complicated than contracts. They contain the standard provisions for roads, fences, ditches, fields, and boundary trees. The legal description of the property is the only additional information needed. Before the deed can be executed, an essential point is to have the seller's signature notarized and in the proper place on the document. Even though it contains occasional changes that most landowners accept, title to the timber passes to the buyer once a properly drawn deed has been delivered.

Sellers should secure their right to control logging through the wording of a contract or deed, but without making the document long or too complicated. In general, longer, more complicated contracts or deeds that have many requirements and restrictions will tend to discourage timber buyers or cause them to submit lower bids. Similarly, too many constraints on logging will increase logging costs, in which case, the seller should be prepared to accept a lower stumpage price. On the other hand, it is the landowner's land and it is reasonable for the landowner to expect to be pleased with the outcome of the sale and harvest. A balance must be struck between having a practical contract that protects you, the landowner, while providing the buyer with sufficient economic incentives to engage in a harvesting operation.

Negotiating the Sale

Whether a landowner handles a timber sale or uses the services of a consulting forester, a significant activity in a sale is contacting possible buyers. Many individuals from employees of processing mills to independent buyers are in the business of buying timber. Thus, most landowners will find several different people who may be interested in their timber sales. Many processing mills have timber procurement foresters who buy timber directly from landowners. They, in turn, will contract with an independent logging firm to cut and haul the purchased timber to their mill.

Many mills also obtain their timber through timber or *wood dealers*. This is particularly true in areas located a good distance from the manufacturing facilities. Wood dealers are independent buyers of standing timber. They can offer an advantage over a single product mill by being able to market many different types of products. Thus, in many cases they may be able to utilize more of the timber on a tract and deliver it to those mills that can best use it. Independent loggers may also purchase timber in the same way as a wood dealer. The major difference is that the logger will harvest the timber personally.

The key to obtaining the highest return for a particular tract of timber is to contact as many different reputable buyers as possible. A list of most buyers operating in a given geographic area can be obtained from one of the public agencies in your county, your consulting forester, or forestry association.

Buyers can best be notified by sending invitations to bid or notifications of availability. Interest often depends on how much factual information is given in the invitation. A county map indicating location and a detailed map of the sale area should be included. In general, four or more weeks should be given buyers to inspect the sale.

Types of Sales

Timber can be sold by two basic methods, *lump-sum* and *per-unit*. Most landowners are not in the timber business and choose to sell their timber using the lump-sum method. The buyer takes title of the standing timber and assumes all responsibility for it, whether all of the timber is cut or not. Payments can be arranged by any method mutually satisfactory to both buyer and seller. The advantages of the lump-sum sales method are ease of administration and no possibility for cheating (since the price is agreed upon up front); also a lump-sum sale places the burden of sorting and merchandising forest products at the landing deck on the buyer rather than the seller. Finally, a lump-sum sale could bring in more revenue than a per-unit sale if the volume has been miscalculated by the high bidder. However, this does not usually occur. In fact, one of the drawbacks to the lump-sum sale is that price is not paid on harvested units. Therefore, a lump-sum sale could end up paying less to the landowner than what would have been paid on a per-unit sale. For many landowners, however, the ease of administering a lump-sum sale outweighs the potential costs of possibly a lower overall revenue.

Per-unit or "pay-as-cut" sales are often conducted where specialty products are sold or where small acreage or low quality timber is present. For example, pay-as-cut sales are typically the preferred contractual arrangements for pulpwood thinnings. Payment is made as timber is cut, based on an agreed price per unit. Units may be in cords, tons, board feet, or piece. It is especially important that the landowner or a landowner's representative such as a consulting forester monitor harvesting operations on a per-unit sale to ensure that what is being harvested is accounted for.

A thorough knowledge of the landowner's tax situation, as well as knowledge regarding what the landowner desires in terms of a sale, and the characteristics of the products harvested from the site are necessary to make a decision on the best method of sale. Knowledge of the buyers in the region is also important. Landowners should consult their tax professional and a consulting forester to make the best decisions especially as tax treatments and regulations change quickly and dramatically. Refer to the Suggested Readings at the end of this chapter for more information.

Owners may choose to sell their timber by sealed bid, auction or oral bid, or by negotiation. Although a trend to sealed bidding is occurring in the South, this may or may not be the best method of sale for an individual or situation. Small sales, sales of specialty products, and sales involving unique circumstances or restrictions may best be handled through negotiations with buyers. Your consulting forester is often the best source of information on your particular situation.

Supervising the Harvest

Most timber buyers are reputable and will follow those contract agreements made with a timber owner. However, an owner and/or forester should ensure that all is well through some type of systematic inspection program. This includes staying in contact with the timber buyer and the contract logger who is cutting the timber, making on-site inspections during the time active logging is occurring, and inspecting the site when cutting is complete.

The owner and/or forester should inspect the timber site regularly to see that all provisions of the timber sale contract are being followed. These inspections should include:

- Checking property lines to see that they remain well marked and all timber felling is within the boundaries.

- Checking to ensure that specifications for complete use of the designated portion of trees are being followed.
- Checking to ensure that those provisions protecting the soil-site quality are being followed by the logging contractor. This includes general condition of the site, the road system, skid trails, and stream bottoms. The final inspection should also include making sure that skid trails, roads, and the site are left with proper drainage structures to eliminate the chance of erosion and the loss of topsoil.
- Checking the compliance of all Best Management Practices (BMPs).
- Ensuring that no unnecessary damage is being done to non-designated trees, fences, ditches, streams, roads, bridges, fields and buildings. A check should also be made during the final inspection to see that any damage is repaired.
- Communicating with the logger to review the provisions of the contract and pointing out the landowner's wishes on-the-ground is always a good idea. This helps avoid misunderstandings and ensures the best job possible.

onsulting Foresters

or most private forest landowners, a timber sale occurs once or twice in a lifetime. Selling timber an be a complicated endeavor when trying to get the most money for the timber and can prove to e costly in terms of lost revenue and damage to the land if mistakes are made. A consulting forester an provide invaluable assistance by representing the best interests of a landowner in all aspects of rest management. A consulting forester can help the landowner devise a management plan, opraise the quantity and value of standing timber, assist with timber sales, and supervise the harvest ensure that the terms of the logging contract are being followed. Following the harvest, a onsulting forester can offer technical assistance regarding site preparation, regeneration of a new op of trees, and timber stand improvement, amongst other services.

The qualifications to become a consulting forester vary from state to state. Most states require resters to pass a written exam and be registered with a state board. Additional qualifications include olding a bachelor's degree or higher in forestry from an accredited university or college, abiding by e Society of American Foresters Code of Ethics, competency in all areas of forest management, opraisal, development, marketing, protection and utilization for the benefit of the general public, nd avoiding conflicts of interest that may compromise the best interests of the forester's client. onsulting foresters may also be members of the Association of Consulting Foresters of America nd/or the Society of American Foresters. Each state forest service or extension service can provide a st of qualified consulting foresters. Appendix E also lists several forestry organizations and ssociations that will have directories of qualified consulting foresters. A telephone directory or eighboring landowner may also be a source of information.

In selecting a consulting forester, request information from three or four of them regarding pecific qualifications, references from previous clients, and fee estimates. Avoid those foresters who are buyers for a single forest products company in order to avoid any possible conflict of interest. After selecting a consulting forester, a contract or written agreement should be signed detailing the services that are to be performed, who will perform them, and the costs of the services. Open channels of communication between the landowner and consultant are critical, and the consultant should be willing to answer any questions and clarifications.

A consulting forester charges a fee for services provided, usually based on an hourly or daily rate, forest acreage, or a contract price based on a percentage of gross revenues from the sale of forest products. The cost of the services can be offset by the increased returns the consulting forester can bring to the landowner. Several surveys and studies concluded that landowners received more incom per acre on average if they utilized a consulting forester prior to harvest. Furthermore, landowners who contracted a consultant's services prior to harvest averaged a 64 percent higher price per board foot and a 120 percent higher projected income stream from future sales due to improved regeneration and stocking.

Summary

Many years are spent bringing a timber crop to maturity. Careful planning is needed by landowners to ensure that they enjoy all of the benefits of these efforts. Understanding what they wish to gain in both the short and long-term is the first step in planning and executing a timber sale. The next steps include determining what volumes are present and what has potential value, what potential buyers are operating in the area, and developing an equitable timber sales contract. The advice from both public agencies and private consulting foresters can be a valuable asset in planning and concluding the timber sale.

Suggested Resources

Readings

- Bardon, Robert E., Rick A. Hamilton, and William E. Gardner. 2003. "Timber Sales: A Planning Guide For Landowners." *North Carolina Cooperative Extension Service Publication #AG-640.* Novembe 2003. 16 p.
- Brinker, R. W. and J. C. Bliss. 1996. "Selling Timber Successfully." *Circular ANR-626.* Alabama Cooperative Extension System, Auburn University. Auburn, AL.
- Dunn, Michael A. 2002. "Marketing and Selling Your Timber." *Louisiana State University Agricultural Center Publication #2893.* December 2002. 8 p.
- "Forestry Handbook, Second Edition." Edited by Karl F. Wenger for the Society of American Foresters. New York: John Wiley & Sons. 1,335 p.

Web Sites

- http://sref.info/
  Southern Regional Extension Forestry Website

- www.timberbuyer.net/topics/logrules.htm
  The Timber Buyers Network
- http://basineducation.uwex.edu/woodland/manage/part3/index.htm
  Wisconsin Woodland Assistance website
- www.woodbin.com/ref/wood/emc.htm
  Woodbin website
- http://www.fltc.net/
  Forest Landowners Tax Council
- http://www.timbertax.org/
  National Timber Tax Website

eview Questions

1. Assume you have a 20 acre tract of pine timber in the southeastern United States. It in located in the pine belt of the South; therefore, it has reasonably good access to markets. A timber cruise of the tract shows that it is predominantly pine sawtimber with some chip-n-saw and pulpwood; however, it is still relatively small sawtimber and this tract is not easily accessible. Now assume you have a second tract of timber, also located in a good area of the South. It is 200 acres of relatively large, mature pine sawtimber and the tract itself is located next to a state highway and less than 30 miles from 3 sawmills.
   a. From a marketing perspective, can you think of some potential advantages that the large tract may have over the small tract?

   b. Assuming that the large tract has a marketing advantage over the small tract, how can the owner of the small tract market the tract to attempt to receive a better price?

Suggested Activities

As an exercise, consider your forestland and assume you are about to make a timber sale. Given wha you have learned here and your own unique circumstances with respect to your land, devise a draft a timber sale contract, including those provisions or stipulations that you deem important. After you have completed your draft contract, have a professional forester review and critique it. This will not only help you think about all the things you need to consider when the time comes for an actual timber sale, but it will also help you to learn what are the important particulars of a contract with respect to your land. You can find examples of timber sale contracts by contacting your state forestry agency, your state cooperative extension service or by typing in "timber sale contract" into your favorite Internet search engine.

Unit VI:

Regulations, Laws, and Politics

ır forest lands, growing and harvesting trees, and hunting and recreation are all subject to rules and ŗs. To successfully manage your forest land and enjoy its benefits, you should understand the basic laws 1 regulations governing the land and its use. The following chapters review these concepts, with the al chapter discussing ways the landowner can influence future changes to these laws and regulations.

apter 13 contains a brief review of the major federal environmental regulations and some discussion their effect on the landowner. Even though you may feel like there are more requirements than one ıld possibly understand, forestry consultants and other professionals are familiar with these regulations 1 will help you comply with them and your local BMP's. You should, however, become familiar with basic regulations to ensure that your silvicultural contracts are followed and the value of your land is t diminished by contractors' taking "shortcuts."

apter 14 reviews your rights as a land owner and the liabilities and obligations you have as a result of ıt ownership. As with all the topics discussed in this book, only the basic concepts are covered. For cific situations, and especially for local considerations, consultation with a professional or in-depth earch will be needed.

apter 15 investigates forest sustainability and some concepts of certification. Currently, there are many ilable certification standards; how each relates to forest sustainability is the subject of ongoing cussions that will continue for years. An important aspect of the certification discussion, particularly wood grown in the US, is that, currently, consumers are not willing to pay a premium for forest oducts in which the wood content originates from certified forests. However, to avoid being surprised er, you should investigate local forest certification activities to understand their impact on your estland plans.

apter 16 suggests ways you can stay informed about the laws, regulations, and policies, and, more portantly, *how you can affect future changes.* Continually staying informed is extremely important since gulations and laws change often, and, it is almost certain that some of the regulation details covered in s book will have changed by the time you read these chapters. State and national forestry ganizations can help you keep up to date on these changes.

tically important is your ability to influence changes to the laws and regulations that affect your use of ır forest lands. Over the last 30 years, forest landowners have let others take the lead and set the enda in forest use policy. By joining with other landowners in state and national organizations like the **rest Landowners Association,** you can promote and influence changes that will encourage healthy est vegetation and nurture diverse and thriving wildlife populations as well as being economically asible.

Chapter 13: Federal Environmental Regulations and the Private Forest Landowner

The earliest rules and regulations governing the 'ests and environment of America date back to 1626. However, many of the early regulations couraging conservation were not enforced due to an abundant resource base. Instead, disposing developing and privately using natural resources was emphasized. It was not until 1919 that ious attention was paid to regulating forestry activities on private lands. The Society of American resters appointed a Committee for the Application of Forestry to investigate ways to prevent forest vastation on privately owned forestland. One of the Committee's recommendations was federal islation for private forestland to prevent mass deforestation from harvesting operations and forest :s.

The majority of current federal legislation affecting private forest landowners was passed during the ·ly 1970's as a result of a groundswell of public support for the environment. Since early Colonial es, federal regulations have stemmed from the state's police power to protect society's public alth, safety, morals and general welfare based on the common law of private nuisance. The nmon law of private nuisance dictates that a landowner may not use a piece of land in a way that tracts from the real property rights of others.

This chapter provides an overview of federal environmental regulations that directly affect forestry ivities on private lands. Environmental quality and public health laws (Clean Water Act; Coastal

Zone Act Reauthorization Amendments, Clean Air Act; and Federal Insecticide, Fungicide, and Rodenticide Act) and the Endangered Species Act always apply to private ownership. Substantive ar procedural statutes (National Environmental Policy Act, Administrative Procedure Act, National Historic Preservation Act, Multiple-Use Sustained-Yield Act, National Forest Management Act, and Wilderness Act) apply to private owners if the private owner receives federal grants, assistance, or permits. Finally, the Rivers and Harbors Appropriation Act of 1899 has the potential to affect privat forestry activities that require a barge terminal.

This body of legislation has been enacted at the federal level but is implemented by the individua states. For further assistance in interpreting these laws as they apply to you, contact your state fores service, a consulting forester, or a county Extension agent.

Federal Water Pollution Control Act of 1972

The Federal Water Pollution Control Act (FWPCA) of 1972 as amended in 1977, 1987 and 1993, most widely known as the Clean Water Act (CWA), was established to combat the severe water quali problems facing the nation after decades of neglect. The basic water quality objective expressed in the CWA is to restore and maintain the chemical, physical and biological integrity of the nation's waters. The CWA set out to accomplish this by improving water quality to "fishable and swimmable" quality by 1983 and by eliminating point source discharges into navigable waters by 1985. Degradation of existing water quality is also prohibited.

Of all the environmental laws passed at the national level, Sections 208 and 404 of the 1972 CWA have impacted forestry activities on private lands the greatest. Section 208 mandated that development, agriculture, mining and forestry activities control nonpoint source (NPS) pollution. Rather than pass state forest practices acts to control NPS pollution from silvicultural activities as suggested by the Environmental Protection Agency (EPA), states developed Best Management Practices (BMPs). Currently, a majority of the states have implemented BMP educational programs for information dissemination and instituted compliance check programs to monitor the successful and effective usage of BMPs.

Section 319 was added to the CWA as an amendment in 1987 to strengthen Section 208. Due to t continuing problem of NPS pollution, Section 319 required the implementation of enhanced planning and control strategies by the states. Section 319 has led to improved and increased use of BMPs in the South, in addition to increased water quality monitoring and enforcement of violators state water quality standards. In order to avoid violating Sections 208 and 319 of the CWA, private forest landowners should implement BMPs if not already doing so; if BMPs have been established, then follow those practices as closely as possible when undertaking any forestry activity.

Section 404 of the CWA requires a permit from the U.S. Army Corps of Engineers prior to dredging or filling any of the nation's waters. The nation's waters are defined as all navigable strean and rivers and wetlands. Following a 1975 court case (Calloway v. Natural Resources Defense Council) attempting to include wetlands in the definition of the nation's waters, the 1977

Amendments to the CWA exempted normal silvicultural activities from obtaining a permit for dredge and fill activities. Final rules issued in 1986 by the Corps of Engineers and the EPA continued this exemption for normal silvicultural activities on pre-existing forestland. However, activities that would enable one to start growing trees are not part of an established operation, and therefore are not exempt from permit requirements. Furthermore, while normal harvesting is exempt, the construction of farm, ranch or forest roads is not exempt. Managed timber production on wetland areas must adhere to state-approved BMPs to be exempt from federal and state permit requirements.

As a private forest landowner, you must demonstrate continuing silvicultural activities to maintain the Section 404 exemption. Having a management plan on file for your tract(s) of forestland and up-to-date written documentation of management practices will help to demonstrate ongoing forestry activities. In addition, timber thinning, periodic harvests, or timber stand improvements would provide other examples of continuing silvicultural management.

Coastal Zone Management Act of 1972

The Coastal Zone Management Act of 1972, as amended in 1990, is applicable to states that border coastal waters. The objective of the act is to provide protection to coastal areas through the development of land and water use plans to preserve, protect, develop and restore or enhance coastal resources. The 1990 amendments expanded the defined coastal zone to include several counties inland from coastal waters. Furthermore, any state with a federally-approved program must implement a plan to control coastal nonpoint source pollution. The EPA is mandated to publish management guidelines to control nonpoint source pollution in an economically practicable manner that use the best available control practices. Finally, any state failing to develop control plans loses federal funding provided via Coastal Zone and Section 319 (Clean Water Act) planning grants.

Clean Air Act of 1970

Although laws were passed in 1955 and 1967 to protect against air pollution, it was not until 1970 and the passage of the comprehensive Clean Air Act (CAA) that any noticeable strides were taken to improve air quality. The CAA was amended in 1977 and 1990. The main objective as described in the 1970 Clean Air Act is to reduce substances in the air that may be harmful to human, animal or plant health by causing growth problems, sickness, mortality or economic losses. The biggest change with the enactment of the CAA in 1970 was that the federal government, via the EPA, took control in setting national air quality standards, and the states became responsible for emission reduction plans, whereas prior to 1970, the states were responsible for setting their own air quality standards.

The part of the CAA that is most applicable to private forest landowners is Section 109. This section requires the EPA to establish national ambient air quality standards (NAAQS) for officially listed pollutants. The six most common types of pollution are carbon monoxide, lead, nitrogen oxides, ozone, particulate matter, and sulfur oxides. Primary NAAQS are the levels of air quality necessary to protect human health. Secondary NAAQS are the levels of air quality required to protect public welfare, meaning anything other than human health such as soils, vegetation and wildlife.

The NAAQS concerning particulate matter are the most relevant to private forest landowners. In 1987 the EPA began measuring particulate matter 10 microns in size or smaller (PM10) and used this standard to regulate prescribed burning and residential wood burning. As mandated by the 1990 Amendments to the CAA, all nonattainment areas were to adhere to new PM10 standards by December 1994. In moderate nonattainment areas, reasonably available control measures (RACM) were to be implemented by December 1993, and serious nonattainment areas required the best available control measures (BACM) by the same deadline to meet the new PM10 standards. Check with your state forest agency, a consulting forester or county extension agent to determine which type of nonattainment area and corresponding control measures, if any, apply when considering prescribed burns.

Federal Insecticide, Fungicide and Rodenticide Act of 1947 and the Federal Environmental Pesticide Control Act of 1972

The Federal Insecticide, Fungicide and Rodenticide Act of 1947 (FIFRA) was established to protect the environment and users from harmful pesticides by requiring the registration of pesticides with the U.S. Department of Agriculture. However, the Secretary of Agriculture had no authority to deny registration or prevent pesticide misuse. The Federal Environmental Pesticide Control Act of 1972 (FEPCA) amended FIFRA and placed the EPA in charge of pesticide regulation and instituted strong federal control over pesticide application.

The EPA is authorized to classify and register the use of most fungicides, herbicides, pesticides and rodenticides. The safety of every registered pesticide is decided based on available scientific evidence, and that chemical is listed regarding applications for which it has been approved. Chemical formulations that are hazardous to the environment can be completely banned, and the only legal uses of chemicals are those applications that have won approval from the EPA. The product label of an herbicide or pesticide is converted into a binding legal document under FIFRA and FEPCA, which makes chemical manufacturers liable for damages when the pesticide is used per label guidelines.

FIFRA, as amended by FEPCA, dictates that all pesticides be registered with the EPA. The pesticide classification system covers general use (pesticides available to the general public) and restricted use (pesticides available only to certified applicators), and distinguishes between classes of applicators. Private applicators can apply pesticides on property they either own or lease, whereas commercial applicators are licensed to apply restricted use pesticides. Both types of applicators must meet minimum competency standards. Finally, FIFRA and FEPCA make it unlawful to misuse pesticides, and they pass the responsibility for enforcement of regulations to the designated state agencies.

Endangered Species Act of 1973

The Endangered Species Act of 1973 (ESA) evolved from the Endangered Species Preservation Act of 1966 and the Endangered Species Conservation Act of 1969 to become the most comprehensive wildlife law to date. The purposes of the ESA are to protect the ecosystems inhabited by endangered and threatened species and to conserve and rebuild endangered and threatened populations of plants and animals to viable levels. The ESA was amended in 1978, 1982 and 1988.

Before a species is federally listed as endangered or threatened, an extensive 16 step scientific process must be undertaken. Once a species is listed, critical habitat is designated for it. Critical habitat is the geographical area physically occupied by the particular species at the time of listing that is necessary for that species survival. If deemed necessary by the Secretary of the Interior, critical habitat can include the historic range of the particular species even if the species does not currently occupy that geographical area at the time of listing.

Section 9 of the ESA prohibits the taking of any endangered or threatened species unless otherwise permitted. As such, Section 9 has wide reaching ramifications due to the "takings" concept, defined not only as physically killing an endangered or threatened species, but expanded to include altering or destroying critical habitat essential to the continued survival of that species. This expanded definition has halted timber harvesting and development on private lands in the southeastern United States where the red-cockaded woodpecker occupies old-growth southern pines. Furthermore, if all plant and animal species that may be endangered or threatened are ever listed and critical habitat is designated and enforced, large tracts of private forestland could be affected from both a management and financial viewpoint.

"Takings" under the ESA is different from eminent domain. "Takings" in the form of reduced or lost economic benefits due to decreased property values or earning potential does not require just compensation. Eminent domain, on the other hand, takes property for public purposes and just compensation is paid. Needless to say, "takings" is a very controversial issue for this reason.

National Environmental Policy Act of 1969

The National Environmental Policy Act of 1969 (NEPA) was the first in a series of environmental laws passed at the national level that impacts the way private timberland is managed. NEPA became effective 1 January, 1970 and was enacted "... to foster and promote the general welfare, to create and maintain conditions under which man and nature can exist in productive harmony, and fulfill the social, economic, and other requirements of present and future generations of Americans." This is accomplished through cooperation between governments at the federal, state and local level as well as other public and private organizations in providing primarily financial, technical and educational assistance. The main thrust of NEPA is the required preparation of an Environmental Impact Statement (EIS) or Environmental Assessment (EA) for any federal action that has the potential to impact the environment.

A number of issues are considered in the preparation of an EIS. First and foremost is the environmental impact of the proposed action. Is the proposed activity a "major federal action that will significantly affect the quality of the human environment"? "Action," as used in the previous sentence, includes physical projects as well as licensing, permitting and certifying. "Federal" includes governmental assistance. The other two criteria ("major" and "significantly affect...") are judgment calls. The second issue to be considered in preparing an EIS is dealing with the adverse environmental effects that can not be avoided. Alternatives to the proposed action are also examined. Can adverse environmental impacts be avoided, and if not, can they be minimized? A fourth step in the EIS process is to balance "local short-term uses" of the environment against "long-term productivity." Finally, the proposed project's "irreversible and irretrievable commitment of resources" is investigated.

The direct application to a private forest landowner, therefore, is that preparation of an EIS would be required of any private landowner who receives federal assistance of any kind and where the potential exists for impacting the environment. Environmental impacts range from building logging roads to actual logging operations. Indirectly, NEPA has been extended to private lands through a test case involving the Tennessee Valley Authority (TVA). The TVA proposed building a wood chipping and barge facility on the Tennessee River. It was determined that an EIS would be required since the barge facility would be built on government-owned land next to the river. One of the alternative proposals required an EIS on all private lands in the approximately six million acre region if a timber harvesting permit is needed and if the private forest landowner wanted to provide wood to the chip mill via the barge loading facility. This test case is an example of how NEPA and the EIS process might impact private forest landowners.

The Administrative Procedure Act (1943) and Subsequent Amendments

Arguably the most important piece of legislation governing federal regulatory agency policy making, the Administrative Procedure Act (APA) of 1946 serves as the foundation for regulatory procedure and establishes the rights of individuals in the regulatory process. Serving to codify, rationalize, unify, and extend haphazardly-applied practice to all federal agencies, the Act helped to clarify the doctrine of procedural due process. The APA has had far reaching effects on the course of economic and social regulation. First, the Act provided a reasonably complete definition of the procedural rights of individuals. The key elements were weak protection through rights of participation in rule-making procedures, but substantially stronger protection in enforcement of these rules. Second, it generated a huge body of law whose political effect is to bias policy in favor of the status quo. By reducing administrative discretion, formal procedures create transactions costs that increase the time and resources needed to change policy. By enhancing the power of the court to overturn agency decisions, formal procedures give organized interests that seek to preserve the status quo a second bite at the apple. Third, notwithstanding the status quo bias, formal procedures expanded the process of agency decision making in such a way that it forced agencies to take into account and respond to the policy preferences of many relevant interests, not just those favored by the president and his appointees.

The National Historic Preservation Act (1966)

Enacted in 1966 and amended in 1970 and 1980, this federal law provides for a National Register of Historic Places to include districts, sites, buildings, structures, and objects significant in American history, architecture, archaeology, and culture. Such places may have national, state or local significance. The act provides funding for the State Historic Preservation Officer and staff to conduct surveys and comprehensive preservation planning. The act establishes standards for state programs and requires states to establish mechanisms for Certified Local Governments to participate in the National Register nomination and funding programs.

Section 106 of the Act requires that federal agencies having direct or indirect jurisdiction over a proposed federal, federally assisted, or federally licensed undertaking, prior to approval of the expenditure of funds or the issuance of a license, take into account the effect of the undertaking on any district, site, building, structure, or object included in or eligible for inclusion in the National Register of Historic Places, and afford the Advisory Council on Historic Preservation (appointed by the President) a reasonable opportunity to comment with regard to the undertaking. Section 110 of the Act directs the heads of all federal agencies to assume responsibility for the preservation of National Register listed or eligible historic properties owned or controlled by their agency. Federal agencies are directed to locate, inventory and nominate properties to the National Register, to exercise caution to protect such properties, and to use such properties to the maximum extent feasible. Other major provisions of Section 110 include documentation of properties adversely affected by federal undertakings, the establishment of trained federal preservation officers in each agency, and the inclusion of the costs of preservation activities as eligible agency project costs.

The Multiple-Use Sustained-Yield Act of 1960 and the National Forest Management Act of 1976

The Multiple-Use and Sustained Yield Act of 1960 is one of the most important laws governing the management of national forests. It recognizes that they are “lands of many uses.” There are over 130 other laws that also apply. Multiple use says that the national forests are to be used for outdoor recreation, range, timber, watershed, and wildlife and fish purposes. Sustained yield means harvest is in balance with growth. The act applies primarily to the 85 million acres of timberlands in the national forests; it does not apply to national parks, which were created to preserve natural features or areas of historical interest. It requires “...harmonious and coordinated management of the various resources...” and not necessarily the combination of uses that will give the greatest dollar return of the greatest unit output. Economic factors are considered, but they do not necessarily control management decisions. All resource values are weighed and trade-offs are made. (It was with the passage of this act that the purpose of our national forests was substantially changed from its original goal of providing timber and clean water for the nation.)

National Forest Management Act (NFMA) of 1976 largely amended the Forest and Rangeland Renewable Resources Planning Act of 1974, which required a national, strategic planning process for renewable resources for the Forest Service, and comprehensive, interdisciplinary land and resource management plans for units of the National Forest System. The law was seen as necessary, because a lawsuit (commonly known as the Monongahela decision) had invalidated most timber practices in the national forests. NFMA substantially enacted detailed guidance for forest planning, particularly in regulating when, where, and how much timber could be harvested and in requiring public involvement in preparing and revising the plans. NFMA also established the Salvage Sale Fund and expanded other Forest Service trust funds and special accounts.

The Wilderness Act (1964)

The Wilderness Act directed the Secretary of the Interior, within 10 years, to review every roadless area of 5,000 or more acres and every roadless island (regardless of size) within National Wildlife Refuge and National Park Systems and to recommend to the President the suitability of each such area or island for inclusion in the National Wilderness Preservation System, with final decisions made by Congress. The Secretary of Agriculture was directed to study and recommend suitable areas in the National Forest System.

The Act provides criteria for determining suitability and establishes restrictions on activities that can be undertaken on a designated area. It authorizes the acceptance of gifts, bequests and contributions in furtherance of the purposes of the Act and requires an annual report at the opening of each session of Congress on the status of the wilderness system. Under authority of this Act, over 25 million acres of land and water in the National Wildlife Refuge System were reviewed. Some 7 million acres in 92 units were found suitable for designation. From these recommendations, as of December 1998, over 6,832,800 acres in 65 units have been established as part of the National Wilderness Preservation System by special Acts of Congress.

The Organic Statutes of the USDA Forest Service, the USDI Fish and Wildlife Service, and the USDI National Park Service

The focus here is on the protective regulatory (PR) policies that affect forestry in the South. Particular emphasis is placed on PR laws and policies protecting and enhancing water quality. Such policies and laws safeguard society by limiting or mandating certain actions by the public and private sectors. They frequently rely on the “stick” of penalties rather than the “carrot” of subsidies or other incentives to accomplish their objectives. Only in a few instances and in limited jurisdictions do PR policies and laws specifically regulate forest management, but all forestland in the South is affected by PR policy. The effects depend on:

- executive or jurisdictional level of the policy (Federal, State, or local)
- forestland ownership category (Federal, State, industrial private, or NIPF)
- owners' management objectives (multiple use, timber/fiber production, or habitat conservation)
- location with respect to urban centers, water bodies, wetlands, and designated critical habitats for endangered species

The National Wildlife Refuge System Administration Act of 1966 amended by the National Wildlife Refuge System Improvement Act of 1997

This act constitutes an "organic act" for the National Wildlife Refuge System. It was amended by the National Wildlife Refuge System Improvement Act of 1997. This law amends and builds upon the National Wildlife Refuge System Administration Act of 1966 to ensure that the National Wildlife Refuge System is managed as a national system of related lands, waters, and interests for the protection and conservation of our Nation's wildlife resources.

The 1966 Act provides guidelines and directives for administration and management of all areas in the system, including "wildlife refuges, areas for the protection and conservation of fish and wildlife that are threatened with extinction, wildlife ranges, game ranges, wildlife management areas, or waterfowl production areas." The Secretary is authorized to permit by regulations the use of any area within the system provided "such uses are compatible with the major purposes for which such areas were established."

Subsequent public law provides that proceeds from disposal of lands in the system acquired with "duck stamp" funds or by donation are to be paid into the Migratory Bird Conservation Fund, and that the Migratory Bird Conservation Commission must be consulted before disposal of any such acquired land. The National Wildlife Refuge System Administration Act Amendments of 1974 requires payment of the fair market value for rights-of-way or other interests granted, with the proceeds deposited into the Migratory Bird Conservation Fund and made available for land acquisition. A 1976 public law 94-215 clarified that acquired lands or interests therein can be exchanged for acquired or public lands.

The Rivers and Harbors Appropriation Act of 1899

Section 9 of this Act, the Rivers and Harbors Appropriation Act of 1899 prohibits the construction of any bridge, dam, dike or causeway over or in navigable waterways of the U.S. without Congressional approval. Administration has been delegated to the Coast Guard. Structures authorized by State legislatures may be built if the affected navigable waters are totally within one State, provided that the plan is approved by the Chief of Engineers and the Secretary of Army.

Under Section 10 of the Act, the building of any wharfs, piers, jetties, and other structures is prohibited without Congressional approval, and excavation or fill within navigable waters requires the approval of the Chief of Engineers. Authority of the Corps of Engineers to issue permits for the discharge of refuse matter into or affecting navigable waters under Section 13 of the 1899 Act was modified by the Federal Water Pollution Control Act Amendments of 1972 that established the National Pollutant Discharge Elimination System Permits. The Fish and Wildlife Coordination Act provides authority for the U.S. Fish and Wildlife Service to review and comment on the effects on fish and wildlife of activities proposed to be undertaken or permitted by the Corps of Engineers. Their concerns include contaminated sediments associated with dredge or fill projects in navigable waters.

Summary

This chapter has reviewed federal environmental legislation and the effects of these policies in relation to forestry activities on private lands. Many of these laws were either enacted or became effective in the early 1970's in response to the declining quality of our environment and the tremendous public support for stricter environmental legislation to protect our natural resources. These laws increase the management responsibilities of private forest landowners as well as the cost of management activities. Nonetheless, it benefits the private landowner to stay abreast of the requirements of and potential amendments to existing laws so compliance can be maintained.

Suggested Resources

Readings

- Coder, K. D. 1994. "Landowners and the Endangered Species Act." *Bulletin 1114.* Georgia Cooperative Extension Service, University of Georgia. Athens, GA.
- Cubbage, F. W. 1993. "Federal Environmental Laws and You." *Forest Farmer, 29th Manual Edition.* 52(3): 15-18.
- Cubbage, F. W., J. O'Laughlin, and C. S. Bullock, III. 1993. "Forest Resource Policy." John Wiley and Sons, New York.
- Office of the Federal Register. 1997. "Code of Federal Regulations, Protection of Environment." US Government Printing Office. Washington, D.C.
- Granskog. J. E., T. Haynes, J. l. Greene, B. A. Doherty, S. Bick, H. L. Haney, Jr., S. O. Moffat , J. Speir and J. J. Spink. 2002. "Chapter 8: policies, regulations and laws. In Southern Forest Resource Assessment." USDA Forest Service, Southern Research Station. *Technical Report GTR SRS-53.* West, D. N. and J. G. Greis (eds.). Asheville, NC. p. 189–223.

Web Sites

- www.srs.fs.usda.gov/sustain/
  Southern Forest Resource Assessment
- www.srs.fs.usda.gov/sustain/
  EPA—Watershed Academy Web
- http://laws.fws.gov/
  FWS—Digest of Federal Resource Laws of Interest to the U.S. Fish and Wildlife Service

Review Questions

1. Why have there been so many federal environmental laws enacted in the past 30 to 35 years?

2. Who generally implements these federal laws, individual states or the federal government?

3. Do environmental quality and public health laws affect all forest landowners?

4. When do substantive and procedural statutes apply to private owners?

Chapter 14:

Private Property Rights and Potential Liabilities

An increasing number of people own private property including timberland. The reasons for ownership are varied; some own land as an investment (appreciating property values), while others reap economic gains from the land (resource marketability such as timber harvesting). Still others own land as a way to improve their quality of life and to escape some of the pressures of everyday life. Whatever the reason, every private property owner has certain rights guaranteed by the Constitution of the United States. With these rights, however, come responsibilities and limits, as defined by the government. There can also be liabilities involved in owning property.

This chapter reviews the elements that define private property, the history of private property rights and the role of the government in relation to private lands. In addition, potential liabilities and the minimization of those liabilities are outlined.

Definition of Private Property Rights

There are four main elements that establish property as private. Private property rights are well defined, exclusive, enforceable and transferable. Well defined property rights make it known where one person's rights begin and another's end. Every tract of private land is defined by a property boundary, and its description is kept on file at the local tax assessor's office. Once the property is well defined, the private property owner enjoys the right to exclude anyone from using resources present on that piece of property. That means anyone wishing to use available resources must first gain permission from the landowner. Exclusivity is effective only if property rights are enforceable. Enforceability is possible because our society and the laws we live by recognize an individual's right to own property. Without society's acceptance of private property rights, it would be difficult to enforce the exclusion of others. Finally, well defined, exclusive and enforceable private property rights enable property to be transferred through the free market process. The free market ensures that the resource, in this case private property, will be allocated in the most efficient manner and to the highest bidder.

History of Private Property Rights

A free market economy underpins the foundation of our country. Historically, it was believed that the most efficient allocation of natural resources was through the marketplace. In order for the market to work, resources had to be privatized and granted a sense of ownership. One of the most valuable resources was land, and the government attempted to transfer property into private hands. The government's main role was to protect private property rights and enforce those rights with a common law doctrine, rather than interfere with the rights of property owners.

With growing economic complexity and population density, government regulation also increased. Public concern for the environment grew through the twentieth century, as did the government's role in regulating actions on private lands. Beginning in the late 1960s and early 1970s, with the passage of numerous federal environmental laws (as outlined in Chapter Thirteen), the government's direction over private lands and the management of natural resources on those lands expanded.

A number of reasons account for the expansion. First and foremost, perhaps, is our country's growing population. With an increasing population base, greater pressure is placed upon our finite land resources. As economic growth takes place, competition for a relatively fixed resource base builds, increasing the scarcity for highly demanded resources. Second, environmental quality is demanded by participants in a prosperous economy. The public demands and expects, among other things, recreational opportunities and a quality environment. Third, a greater percentage of our population is educated today than ever before. With a more educated public, changes in attitudes regarding our environment have taken place which have increased the call for greater government involvement in protecting the quality of our nation's natural resources.

The Government's Role in Private Property Rights

A landowner who possesses rights to property which are well defined, exclusive, enforceable and transferable is called a fee simple owner. While a fee simple owner has relatively free control over the management of privately owned land and the resources present, the government limits private property rights with certain rights of its own exercised on behalf of society and the general public. The government retains the right to tax private property, the right to take private property for public use (eminent domain), the right to regulate actions on private property (police power) and the right to escheat private property.

The government, since its establishment, has taxed private property in various ways. At the federal level, income tax and estate and gift taxes are three tax treatments that directly affect private forest landowners. If property is inherited, the landowner must pay an estate tax. At the state and local level, property tax must be paid, and in most states, state income tax is collected. Local governments usually use property tax as the primary source of funding for schools, roads, police and fire protection and other publicly provided services.

The federal government has the right to take private property to further a public purpose, commonly known as eminent domain. A frequent example of eminent domain is the confiscation of private property to build a highway. If the government exercises its right to eminent domain, a just compensation payment must be provided to the landowner as required by the Fifth Amendment to the United States Constitution. Likewise, the Fourteenth Amendment to the Constitution prevents individual states from confiscating private property without due process of law. Recently there have been instances where governments are taking property for private uses, and these cases are being ajudicated in the courts.

No landowner has an absolute right of use for a piece of property. A landowner must use property in a manner that preserves the public's health or welfare and the interests of neighboring landowners or the community as a whole. The limits to landowner rights are dictated by the state's police power, the legal right of the government to regulate actions on private property in order to ensure the public's health, safety, morality and general welfare. Examples of environmental regulations that directly affect private forest landowners are the Clean Water Act, the Clean Air Act and the Endangered Species Act (see Chapter Thirteen).

The last formal right of the government regarding private property is the right to escheat. This enables the government to take possession of private lands when owners with no known heirs die without a will.

The Takings Issue and Private Property Rights

While the physical taking of private property for public use requires just compensation, the concept of regulatory takings is an ambiguous one. A regulatory taking is defined as a governmental regulation that restricts private land use to the point where economic gains are either completely removed or severely impaired. A 1922 United States Supreme Court decision (Pennsylvania Coal Company v. Mahon) ruled that government regulations constituted a taking even though no physical taking had occurred. A law was passed by the State of Pennsylvania prohibiting coal mining where mining activities would cause structures or streets to fall into the mined areas. The Court found this regulation to violate the Fifth Amendment since the landowners (Pennsylvania Coal Company) were denied economic rights to the land due to overly burdensome regulations. The United States Supreme Court heard a case (Keystone Bituminous Coal Association v. DeBenedictus) in 1987 that was similar in nature and found that a regulatory taking did not occur since the mining regulations that had been enacted did not deny the coal companies all "economically viable use" of their land.

Although regulatory takings cases have been heard by the courts since 1922, the aforementioned takings cases exemplify the difficulty of establishing hard and fast rules to decide when a regulatory taking has occurred. There are, however, some general questions the courts consider. The first concerns the economic impact of the regulation on the landowner. In recent years, the courts have attempted to determine if the landowner is left with a reasonable economic use of the land after the regulation goes into effect. What determines a reasonable economic use is certainly open for debate and is decided on a case by case basis. Denying the landowner the "highest and best use" of a piece of property fails to constitute a taking in the court's eyes. The second question often asked is how valid is the public purpose the regulation aims to promote. The courts usually bow to the opinion of public officials when deciding what is a legitimate public purpose for regulation. The definition of what is a valid public purpose for land-use and environmental regulations has been expanding recently to include open-space, agricultural land conservation and environmentally sensitive areas like wetlands and floodplains. The last in this series of questions looked at by the courts is the character of the government action. For example, is a developer required to provide a certain amount of open space if the development being constructed generates the need for that open space? Generally speaking, the courts have allowed governments to place conditions on newly developed lands as long as the correlation can be proven that there is sufficient need created by the project and there is a sufficient amount of land available.

What does all of this mean to the private forest landowner? Any environmental regulation that diminishes the economic gains received from private lands most likely will not qualify as a regulatory taking since very few regulations completely extinguish all economic benefits on private forestland. A case in point is the presence of red-cockaded woodpeckers (RCW), an endangered species, in old-growth southern pines found on private lands. While many timber harvests have been delayed or prohibited due to the RCW's presence and the requirement to preserve critical habitat as stated in the Endangered Species Act of 1973, private landowners have received no just compensation since no violation of the Fifth Amendment has been found. The balance between property rights and public welfare rights may be swinging toward property rights as vacancies on court benches are filled in recent years with judges sympathetic to private landowners. With fewer and fewer people being involved in the production of food and fiber, public sympathies are more likely to sway the other way.

Potential Liabilities of Land Ownership

Along with certain rights and responsibilities as a private landowner come potential liabilities. Increasing demands are being placed on private lands to provide for hunting, recreational and educational use. For these and other purposes, users may enter private property legally or illegally. Three types of land users are recognized by law—trespassers, licensees and invitees.

Trespassers are those who enter private property without the permission of the landowner, and thus are afforded the least amount of protection under the law. To avoid liability, the landowner must not willfully or wantonly injure the trespasser. Where children are involved, the charge of trespassing may be overlooked by the courts if the principle of "attractive nuisance" applies. It is the landowner's responsibility in cases of attractive nuisance to prevent access by children to a site or condition on the property that the owner recognizes as possibly causing harm to a child. A pond or open pit is an example of an attractive nuisance.

Licensees enter private property with the landowner's permission, but they provide neither direct nor indirect benefits to the landowner. A licensee is provided greater legal protection than a trespasser but less than an invitee. When granting permission for access to private lands, the landowner should treat the licensee as if they were an invitee.

Invitees enter private property with the landowner's consent and knowledge and the arrangement is to each party's mutual benefit. Direct benefits may be given as in the case of an invitee paying the landowner for hunting privileges. The greatest protection under the law is provided to the invitee; consequently the invitee poses the greatest liability for the landowner. If permission is implied, the status of a trespasser may change to that of an invitee.

Each person, whether a licensee or an invitee, should be informed of possibly hazardous areas present on private property. Examples of dangerous areas include large holes, ponds and streams, and dead or dying trees. Furthermore, the locations of loggers or hunters in the area should be provided.

Reducing Potential Liability

There are a number of steps that can be taken to minimize landowner liability. Purchasing liability insurance can protect landowner assets and properly posting property signs can offer protection from trespassers. Signs should be at least 10-by-12 inches in size and posted no more than 200 yards apart (check on the regulations in your state). At the very minimum, at least one sign should be posted on each side and at each corner of the property. Post signs out of reach to prevent vandalism, and replace any worn, damaged or stolen signs. Erect gates across any private entrance to discourage entry. Any person who uses private land must be warned of potential hazards that may be present. Failure to warn users may cause the landowner to be found negligent or reckless if the land user is injured. Removing or fencing in attractive nuisances can reduce the risk of children entering private property and getting injured. Finally, a well written legal contract between the landowner and land user can protect against liability through negligence claims. Hunting lease arrangements should require both hunting and general liability coverage. Do contact a lawyer to ensure protecting yourself against liability if using a contract.

Summary

The rights of private property owners have changed from exercising absolute dominion over their land and available resources to a larger governmental involvement in the management of private lands. The number of environmental regulations affecting private property has increased in the last several decades; however the land may have increased in value and the satisfaction derived from property ownership may have blossomed. For example, the presence of an endangered species may reduce the economic benefits received for one landowner, while another landowner would be willing to pay more for a tract of land simply due to the pleasure gained from viewing or having such a rare and unique plant or animal on their land.

Whether the balance will tilt more toward government involvement or private property rights in the future is difficult to foresee. Since societal values drive public policy, one can be guaranteed that the changes will evolve slowly in either direction. What is certain is that as our nation's population increases, greater demands will be placed upon private lands and the resources they hold. Private landowners collectively hold the key to meeting those demands. Acting responsibly as owners is the surest way to preserve our property rights.

Suggested Resources

Readings

- Barlow, R. 1990. "Who Owns Your Land?" Southern Rural Development Center. *Publication No. 126.*
- Cubbage, F. W., J. O'Laughlin, and C. S. Bullock, III. 1993. "Forest Resource Policy." John Wiley and Sons, New York.
- Jones, E. J., R. A. Hamilton, M. A. Megalos, and T. Feitshans. 1996. "Land ownership, liability, and the law in North Carolina." Publication WON-21. North Carolina Cooperative Extension Service, North Carolina State University. Raleigh, NC.
- Megalos, M. A., R. A. Hamilton and T. Feitshans. 1997. "Maintaining forest property boundaries." *Publication WON-35.* North Carolina Cooperative Extension Service, North Carolina State University. Raleigh, NC.

Chapter 15:
Forest Sustainability and Certification

Sustainability is a big, broad general idea bringing attention to the conflict between human wants and needs and natural resource limits. It involves the relationship of ecological, economic and social aspects to renewable and non-renewable natural resource availability. When applied to forestry it is an expansion of the early principle of sustained yield. Sustained yield addressed only the narrow and more precisely defined balancing of harvest and growth of merchantable timber. The objective of that principle was to prevent over-cutting the forest but with no other acknowledgement of conditions for the broader picture of ecological, social or economic conditions.

"The concept of sustainability remains amorphous, and some consider it more a philosophy than a set of management practices or observable condition of the forest." This description is from The Forest History Society's publication "Forest Sustainability" by Donald W. Floyd, who also stated "...we are still in the process of debating and defining the meanings of sustainability." There is no agreement on whether or how sustainability is applicable to individual ownerships.

Discussions of sustainability began appearing widely in the early 1980s. The concept of certification in forestry appeared in the late 1980s, perhaps in response to concerns about sustainability in general and tropical deforestation in particular. Certification of sustainability was not possible without a precise definition. It has evolved to the idea of setting standards to improve forest management and then declaring forests managed under those standards, and/or the products subsequently produced, as "certified." Thus, the primary purpose of certification is to at least imply, if not demonstrate, that forests are by some standard, well managed and presumably therefore, sustainable. The first promotion of certification was based on the assumption that educated consumers would pay a premium for products from well managed forests. This has not proven true, and recent efforts have been to convince sellers (retailers) and growers of forests that certified wood products will have greater market access and market share. The jury is still out on the effectiveness of this strategy.

Today in the U.S. there are several certification systems (also called programs or schemes); others exist only in Europe or Canada, with a reported 50 or more certification systems worldwide. A small "industry" has developed around the development, marketing and implementation of such schemes. Funding has come from private foundations with an environmental agenda, or from elements of the forest products business. The specific standards and general emphasis of the systems vary according to the goals of the organizing entity. The requirements of the Forest Stewardship Council (FSC) limit the use of herbicides and plantations along with advocating numerous social policies. The Sustainable Forestry Initiative (SFI) reflects the interest of industrial forest owners in encouraging timber production. Compared to forestry practices of previous decades, all certification programs place more emphasis on environmental goals and involve social aspects of forest resource management.

Primarily for reasons of economics, certification schemes in the U.S. have, for the most part, been adapted to and adopted by large forest ownerships. Because costs are involved and financial rewards are, at this point, speculative, few of the smaller non-industrial private forest owners (NIPF) have been motivated to become certified. In view of the generally good condition of forests in the United States and the fact that there has been a net gain in forest area over the last 80 years in the United States, some question the need or justification for massive efforts to certify our forests. On the other hand, retailers pressured by environmental organizations have adopted policies favoring certified forest products, even though studies have shown few consumers are willing to pay more for certified products. One description of today's situation comes from Clifford F. Schneider–Vice President of Sustainable Forestry for MeadWestvaco:

> "I don't know if certification will exist in 100 years, but I do know it is here now. And people need it as a way of addressing opposition to managed forests by certain groups in our society. For the present, many agree that certification is a useful tool to communicate with customers, the general public, government agencies, and our employees. It helps to develop messages about forest practices, illegal logging, and biodiversity conservation, and it demonstrates we are doing something about these concerns."

Mr. Schneider believes the issue is not technical but societal and "at the core" political, and he believes certification is an effective tool to combat the notion that harvesting wood is wrong. Others express fears that certification is a tool to control management practices, runs counter to free enterprise and infringes on the private property rights of landowners. In any case, certification continues to evolve, and it behooves landowners to keep informed about the various programs and possible benefits.

A listing of the eleven *SFI Objectives* and ten *FSC Principles* (as paraphrased in a paper by the Southern Center for Sustainable Forests) gives a flavor of the differences that can exist between programs.

Sustainable Forestry Initiative (SFI) objectives

1. Broaden the implementation of sustainable forestry, including having a written policy or program, providing funding for forest research, providing recreation and education opportunities, and ensuring that long-term harvest levels are sustainable.
2. Ensure long-term forest productivity and reforestation, through reforestation by natural or planted methods within two years; promote state-level reporting of the overall rates of reforestation success and afforestation; use chemicals prudently and follow BMPs; implement management practices to protect and maintain soil productivity; protect forests from damaging insects, diseases, or fires; and use genetically improved material with sound scientific methods.
3. Protect the water quality by using BMPs developed under EPA-approved state water quality programs and meet or exceed all state water quality laws; develop, implement and document riparian protection measures; provide funding for water quality research; require BMP training for company employees in woodlands and procurement; and encourage training for forest management and harvesting contractors.
4. Manage the quality and distribution of wildlife habitat and contribute to the conservation of biological diversity by having programs and plans to promote habitat diversity at the stand and landscape level; fund research; apply research and technology and practical experience in wildlife and biodiversity management.
5. Manage the visual impacts of harvesting and other forest operations, through planning and design; manage the size of clearcuts with an average size not to exceed 120 acres; adopting a 5 year (5 feet) green-up requirement before adjacent areas may be clearcut; and vary harvest units to promote diversity.
6. Manage lands of ecological, historical and geological significance carefully.
7. Promote the efficient use of forest resources by minimizing waste and ensuring efficient utilization of forest resources.
8. Cooperate with forest landowners, wood producers and consulting foresters, by encouraging use of BMPs, providing environmental and economic information about BMPs and working closely with state logging and/or forestry associations and agencies.
9. Publicly report progress in fulfilling program participants' commitment to sustainable forestry.

10. Provide opportunities for the public and forestry community to participate in the commitment to sustainable forestry.
11. Promote continual improvement in the practice of sustainable forestry and monitor, measure and report performance in achieving the commitment to sustainable forestry.

Forest Stewardship Council (FSC) Principles

1. *Compliance with Laws and FSC Principles.* Forest management shall respect all applicable laws of the country in which they occur, and international treaties and agreements to which the country is a signatory, and comply with all FSC Principles and Criteria.
2. *Tenure and Use Rights and Responsibilities.* Long-term tenure and use rights to the land and forest resources shall be clearly defined, documented, and legally established.
3. *Indigenous Peoples' Rights.* The legal and customary rights of indigenous peoples to own, use and manage their lands, territories and resources shall be recognized and respected.
4. *Community Relations and Workers' Rights.* Forest management operations shall maintain or enhance long-term social and economic well being of forest workers and local communities.
5. *Benefits From the Forest.* Forest management operations shall encourage the efficient use of the forest's multiple products and services to ensure economic viability and a wide range of environmental and social benefits.
6. *Environmental Impact.* Forest management shall conserve biological diversity and its associate values, water resources, soils, and unique and fragile ecosystems and landscapes, and by so doing maintain the ecological functions and integrity of the forest.
7. *Management Plan.* A management plan—appropriate to the scale and intensity of the operations—shall be written, implemented, and kept up to date. The long-term objectives of management and the means of achieving them shall be clearly stated.
8. *Monitoring and Assessment.* Monitoring shall be conducted—appropriate to the scale and intensity of forest management—to assess the condition of the forest, yields of forest products, chain of custody, management activities and their social and environmental impacts.
9. *Maintenance of High Conservation Value Forests.* Management activities in high conservation value forests shall maintain or enhance the attributes, which define such forests. Decisions regarding high conservation value forest shall always be considered in the context of a precautionary approach.
10. *Plantations.* Plantations shall be planned and managed in accordance with Principle and Criteria 1–9, and Principle 10 and its Criteria. While plantations can provide an array of social and economic benefits, and can contribute to satisfying the world's needs for forest products they should complement the management of, reduce pressures on, and promote the restoration and conservation of natural resources.

Many companies and individuals seeking and achieving certification have found their practices agreed closely with requirements of the chosen program. The time consuming and therefore costly part of the effort was in establishing and maintaining an administrative system to record their compliance. This is necessary because most systems require regular independent audits of the forest conditions and management activities. Documentation suitable for auditing requires a well designed system of tracking intentions and accomplishments. Creation and maintenance of a system for keeping such records is a major hurdle in achieving certification.

It is apparent FSC and SFI were not designed primarily for non-industrial private forest owners, since many of the above requirements are not suitable for the 40 acre or even the 1,000 acre property. To address this, some programs have been tailored for the highly variable and less structured ownerships of non-industrial private forest owners. An example is the Standard of Sustainability developed for implementation in 2004 by the American Forest Foundation for certification of a group. Group certification is designed to be available to forest owner organizations which can include consulting foresters, associations, landowner cooperatives, etc. Following an initial audit, third party recertification of the group is required every five years. Another program structured for non-industrial private forest owners was developed when the National Woodland Owners Association joined with members of the Association of Consulting Foresters, to create Green Tag Forestry certification under the National Forestry Association.

In summary, forest certification is well established in the U.S. and has been designed for, and accepted by, many large private and public ownerships. The costs of implementation and the lack of market advantage or increased returns for products sold from certified forests have resulted in limited acceptance by non-industrial private forest owners. Aspects of certification will continue to evolve, and landowners and advisors to landowners should continue to monitor developments in this area.

Suggested Resources

Readings

- Floyd, Donald W. 2002 "Forest Sustainability; the History, the Challenge, the Promise." Durham, North Carolina: Forest History Society.
- Cubbage, Frederick, et.al. 2002 "Implementing Forest Certification in North Carolina: Systems, Costs, and Forest Management Implications; Southern Center for Sustainable Forests"; *Proceedings, 2002 Southern Forest Economics Workshop.*
- Sample, V, Alaric 2004 "Forest Plantations as Components in a Global Biodiversity Conservation Strategy: The Role of Developed, Temperate Forest Countries (Part I)." *The Pinchot Letter* Vol. 9, No. 1. Washington, D.C.: The Pinchot Institute for Conservation.

- Schneider, Clifford F. 2003 "The Political Dynamics of Sustainable Forestry", Presented at the conference "Practical Sustainable Forestry and the Marketplace: Making it Work." Adams Mark Hotel, Jacksonville Florida, June 30, 2003.

Web Sites

- www.treefarmsystem.org
  American Tree Farm System
- www.fscus.org
  Forest Stewardship Council
- www.woodlandowners.org
  Green Tag Forestry
- www.env.duke.edu/scsf
  Southern Center for Sustainable Forests
- www.afandpa.org
  Sustainable Forestry Initiative

Chapter 16:
Being Informed, Taking Action

Since the middle of the 20th century Americans have been moving to the cities and away from the farms. The promise of better jobs and better futures for their families created an America today that is detached from the land and dependant on those of us who continue to manage our lands for food and fiber production. Forestland in general has steadily increased over the past 50 years due to decreased use of the land for agriculture, government incentive programs aimed to protect erodeable lands, and strong markets for forest products. The overall result of these events has brought about a misunderstanding in the general public about the management and benefits of healthy forest. Healthy forests not only provide our growing nation with a renewable fiber supply but they have key environmental benefits such as soil protection, increased air and water quality and habitat for both game and non-game wildlife species.

With all of these factors it would seem that the public would want to encourage private forest landowners to continue their practices. However, the misunderstanding that exists by the public causes many knee jerk reactions in the form of legislation and regulations aimed at restricting landowners rights to properly manage their forest. There has also been an increase in the number of activist groups whose goal is the preservation of all forestlands. These often misinformed but well funded groups make their mark by playing on the fears of many Americans with what we refer to as the "chicken little" theory. Unfortunately, their stories tend to dominate media coverage of these issues and the true, good, story gets left behind. This is the reason it is so important for private landowners to get actively involved and become more informed about the risks they face from their local, state and federal governments and from regulatory agencies.

This chapter will detail a three-step process that all landowners should take to properly ensure that they can continue managing their forest today and for future generations. The first step is to *get involved* with the organizations that work to support you everyday. The second step is to *become informed* about current laws and regulations and about the viewpoint of your elected officials. The final step is to *tell your story* about your forestland and the benefits it provides to as many groups and individuals as possible.

By following these steps landowners are helping to ensure a favorable business and political climate for the future of their forestland.

Getting Involved

One of the best ways to become better informed about the issues that affect you and your land is to get involved with one or more of the organizations that were formed to serve you. There are organizations and associations which can provide you with educational material, representation on state and federal issues, local networking opportunities, and issue-specific activities. All of these groups have something to offer and you should choose those that best fit your initial interest and involvement level. Once you become a member or supporter of your particular organization or association you will start receiving information about the issues and activities that they are working on. With a very small investment you are now ready to learn more about the issues of importance to you, and you will be able to become active on these issues if necessary.

Numerous possibilities are available to you in becoming active with an issue. The following options serve as an excellent starting point. Choose the alternative or combination of options that you feel the most comfortable with and that is best suited to your level of commitment.

Communicating with Legislators

As a citizen and activist speaking on behalf of private woodland owners, you have the inherent opportunity to establish effective communications with your lawmakers. This is where involvement with associations can become very helpful. Associations can provide you with talking points, history of a particular issue, contact information and proper language and procedure when addressing lawmakers and agency employees. Your contacts can have an impact on the outcome of legislation relating to the use and preservation of America's forestland, and possibly your future. These contacts will be much more successful, however, if you make an effort to develop a personal relationship with your elected officials and their staffs. Coordinating your efforts with other groups and organizations can make your efforts even more successful. When was the last time you contacted your elected officials? Where do they stand on the issues of importance to the future of your investment?

To become acquainted with your legislators, you will need both patience and persistence. Most lawmakers and their staffs want to know as many of their constituents as possible as this allows representatives the chance to become better familiarized with issues that are important to them. To attain this end, the majority of lawmakers will usually schedule forums or other speaking occasions designed to give the public an opportunity to discuss relevant issues. These meetings can afford you an excellent opportunity to become better acquainted with your representatives. Keep in mind, however, that your legislators represent many voters and have various viewpoints to consider, so patience and persistence are essential in effectively conveying your concerns to them.

In addition to personal meetings, letter writing is a very effective means of political communication. Considering that much of a lawmaker's time is spent in Washington, D.C. or the state capital, most of your communications will probably be written. The following tips are some good guidelines to consider when writing to your legislator.

- Be sure to address the letter correctly. Consult your local directory, association government affairs director, library, or Chamber of Commerce for the state and federal addresses for your district. The Internet is also a very useful tool that can locate all of your elected officials by your zip + four code for your home.
- State your purpose for writing. It is best to discuss briefly and concisely just one issue per letter. Also identify the issue clearly by the specific bill number and title, if available.
- Acknowledge your lawmaker's stand on a particular issue if known. Be sure to support this position if it agrees with your own, and conversely, challenge this view by stating your beliefs and concerns when it conflicts with your position. In either case, be very courteous and respectful in all of your comments.
- Offer constructive alternatives to a problem. Keep in mind, however, that these suggestions should be reasonable and realistic possibilities.
- Avoid form letters that contain stereotypical phrases or ideas unless you are responding on behalf of a grassroots action where volume of letters is the key. Always try to write in words that accurately reflect your own personality and style.
- Remember to write your legislator regarding actions that are deserving of approval or thanks. These types of letters can help to create a favorable receptivity for later communications.
- Since the tragedy of 9/11 the most effective way to send your letter is via e-mail or fax. Every elected official can be contacted by both of these methods, but faxes are preferred. It is still a good idea to follow your letter with a phone call to ensure it was received.

Whether you choose to meet your representatives personally or correspond with them by letter, you should have a good idea of their background. Your contacts will be much more meaningful and productive if you know their past voting records, issue priorities, what committees they serve on, positions on relevant issues, etc. This information will enable you to discuss your concerns on an informed level, thus affording you a greater opportunity for influence and possible action.

You should also make a point of getting to know your legislator's staff members. In many instances, staff members may be more accessible and approachable than your representative. Staff aides also play a key role in bringing issues to your lawmakers' attention and relaying constituent concerns. So although you may not be able to communicate directly with your elected officials, correspondence with their staff members may be just as productive.

Communication with your lawmakers and their staff can be an extremely effective means of becoming involved. However you decide to do so, it is essential that you continue to follow up on your original correspondence. Many people make the initial step of meeting or writing their representative, but then make the mistake of failing to follow-up with further communications. It is usually only through continual and persistent communication that your efforts will be rewarded.

Groups and Organizations

Joining a group or organization that shares your concerns is another excellent way to get involved. This will not only help you to stay informed on many issues, it will also maximize your influence and actions. A coordinated and organized movement of many people will usually have more influence and power than one person acting alone. This is not to say that your individual actions are meaningless, rather the united voices of many people are generally heard more clearly over the voice of one person speaking alone.

You will discover a myriad of groups and organizations when researching possible groups to join. Each of these associations represents a wide range of issues and ideals. Try to choose a group that is actively involved in promoting your concerns and beliefs. You will probably find that your local, grassroots organizations best represent your views as they generally focus on local issues and concerns. However, some local groups have become increasingly influential and active in national issues as well. The many national groups and organizations also provide comprehensive coverage and leadership on various issues faced by private woodland owners. Consult the "supplemental readings" and "Internet addresses" sections at the end of this chapter to gain further information on groups and associations that interest you.

Community Involvement

As a woodland owner, some of your management activities on your land will undoubtedly be questioned by some people in your community. This is especially true if you manage your forestland for timber production; several operations, such as clearcutting, typically alter the appearance and composition of your forests drastically. Although these practices are useful and necessary management tools, others in your community may view these actions in a disapproving light. Most forest landowners will derive maximum coverage by joining both the Forest Landowners Association and their state forestry association. Your involvement in these associations will give you the necessary information to discuss the management of your land from a scientific and practical point of view. This is not making excuses or pacifying the community but giving them the reasons why certain practices are used in the management of healthy forest. Without the proper information to discuss your practices the potential for increased misunderstanding and alienation of your neighbors may occur. For a list of contacts, see the end of this chapter and Appendix E.

Getting Informed

Once you have made the decision to get involved, the first, and undoubtedly most crucial step, is to become informed. Associations can provide you with a voice on issues but they cannot go it alone. It is important to read the material they produce about particular issues and the day to day practices of landowners. Associations can also provide you with the opposing side of a particular topic. This is essential because you need to become well informed with both sides of an issue to communicate competently. Furthermore, this knowledge will enable you to anticipate the opposition's key arguments and refute these elements in an insightful and effective manner.

To attain this end, you can select from numerous sources to become better informed. The following sources of information are by no means the only options available to you. New methods of gathering information are constantly being developed.

Media Resources

Perhaps the largest sources of information are the many types of media. Be sure to note that some media resources are more dependable than others, and newspapers in general are often very biased and unreliable. The sources listed below should be easily attainable to you at your local library, Internet, or cooperative extension office.

Newsprint and Other Periodicals

If you are trying to locate information on current topics, the printed media is usually a good source. Newspaper articles generally follow current and particularly controversial stories. They will also provide you with some sound background information and will usually attempt to present both sides of an issue. It is important to remember, however, that almost every newspaper will report a story with some degree of bias. Keep this in mind, especially if you are using this as the primary source for researching a particular issue.

In addition to newsprint, periodicals are another useful source of information. Most national magazines and periodicals will give you an in-depth look into a particular subject and may provide you with a clearer understanding of key concepts. Many associations and groups also release publications on relevant topics being discussed in the industry. This is especially true in forestry where numerous groups, such as the Society of American Foresters, Forest Landowners Association, and most state forestry associations all publish reports dealing with a wide variety of topics concerning the many viewpoints associated with the forest industry. For information specific to Forest Landowners, you may wish to subscribe to the FLA Washington Update, Forest Landowner Tax Council Newsletter, and the Forest Landowners Magazine. These types of publications can be useful in providing technical information and key statistics relating to a particular issue that often cannot be found elsewhere.

Television

The tremendous power and scope of television make it a serious force in influencing the thoughts of almost every American on a wide range of issues. One cannot help but be influenced and informed by the vast amounts of information that are constantly being transmitted into our lives. This exceptional exposure allows you to get a good idea of what is happening with a variety of topics and discover some of the more relevant points associated with these issues. Television is also an excellent way to keep you abreast of current issues that can affect private landowners, such as taxes, environmental regulations, and economic forecasts.

Conferences and Workshops

Attending conferences and workshops on issues faced by private landowners is another effective way to become informed. These meetings can be an exceptional place to gain exposure to various concepts and opinions on all types of issues. They can also afford you an opportunity to establish important contacts while interacting with others who share similar concerns as yourself. Again, make it a point to attend a conference or workshop dealing with a subject with which you may not agree. The experience will help you to become more familiar with both sides of an issue.

The Internet

One of the most extensive and fascinating sources of information available to you is the Internet. It is capable of providing an enormous amount of information from all over the world. The Internet has revolutionized the way in which information is distributed and accessed. You can obtain information on virtually any topic by accessing a worldwide resource data bank via your computer in an efficient and relatively cheap manner. You may also be able to make contacts over the Internet with other people who are interested in the same issues as yourself, an especially important consideration if you are trying to gain support for a particular matter.

Similar to the information presented by the media, you should take some of the information found on the Internet with some caution. There is the potential for useless information on the Internet, some of which may be heavily biased, or even incorrect. It is wise to verify this type of information, especially if it comes from an unknown source.

No matter how you go about obtaining information about a particular issue, the key is to utilize as many sources as possible and then make your own determination about where you stand. Not only will this give you exposure to many different viewpoints, it will also help to eliminate some of the bias that is typically found in many of today's widely used informational resources.

Telling your Story

"The squeaky wheel gets the grease." This is also true when it comes to issues, legislation, and regulation. Many of the issues are primarily controlled by those who stand up and make their voice heard, by those who speak out and act on their conviction. That is the heart of what activism is all about — someone who will take direct action in support of or in opposition to an issue. In almost every case, the individuals and groups who actively get involved are the ones who actually make a difference. Your membership in local and national associations that represent your interest will help you become informed, and they will speak on your behalf about these issues, but you cannot let it stop there. Your personnel testimony and stories about why you own land and your aspirations for the future of that land are so very important for the general public to understand our management. The more people understand the deep love that a landowner feels for a piece of property the more they come to realize that private landowners are the best stewards of natural resources and that any future regulation is unnecessary. There are many places where your story needs to be heard.

Involvement in your local community is essential for woodland owners. You can avoid many punitive local ordinances by establishing a good working relationship with your community leaders. A good relationship can provide you the opportunity to explain to the people in your community the purpose behind some of your management practices, like clearcutting, for example. Once they understand these actions, they may be less likely to object to them.

An excellent way to begin developing a healthy relationship with your community is by establishing a demonstration forest on your woodland. A demonstration forest showcases the major techniques commonly used in sound forest management. These forests might include areas that highlight clearcuts, shelterwood cuts, timber stand improvements, regeneration strategies, wildlife habitats, and stream management zones. They also frequently include a comparison of managed and unmanaged areas so visitors can actually see the benefits of forest management. That is the primary goal of demonstration forests.

By observing firsthand the practices involved with forest management, people may begin to see how the advantages of actively managing a forest far outweigh most of the disadvantages. This understanding can go a long way in changing many of the negative attitudes associated with forestry in your community.

Demonstration forests can also afford you the opportunity to improve your public speaking skills, another important tool in community involvement. Taking advantage of as many speaking opportunities as possible is another effective way of building a strong relationship with your community. Even if you own just a few acres, speaking about your knowledge of forest management can help to enlighten your community about the benefits of forestry while strengthening your standing in the community.

Summary

Getting informed and taking action is essential as many landowners throughout the southeast are facing strict government regulations and growing public pressures. It is up to woodland owners like you, owners who have taken steps to improve their land and livelihood, to take the next step and face these challenges by becoming an active voice in the political arena, in organizations, and in your community. You who actively get involved, who stand up and make your voice heard, are the ones who will change the future and make the difference.

Review Questions

1. Why is activism important for private woodland owners?

2. Why should you be informed on the issues before you become an activist?

3. What is the importance of communicating with your elected officials?

4. How have you become involved with your community?

5. What are some other methods of becoming involved?

Suggested Resources

Web Sites

- www.forestlandowners.com
  Forest Landowners Association
- www.congress.org
  Find your member of Congress
- www.thomas.loc.gov
  View federal bills

- www.bugwood.org/silviculture/forestryschools.html
- www.sfrc.ufl.edu/otherschools.html
- http://forestry.about.com/od/forestryeducation/
  Forestry Schools Lists
- www.stateforesters.org/SFlinks.html
  National Association of State Foresters
- www.timbertax.org
  Timber Tax Information
- www.epa.gov/OWOW/TMDL/INDEX.html
  EPA site on Total Maximum Daily Load (TMDLs)
- www.nrcs.usda.gov/
  Natural Resource Conservation Service
- www.swcs.org/
  Soil and Water Conservation Service
- www.deathtax.org
  Repeal of the Federal Death Tax
- www.beconstructive.com
  Wood Promotion Network

Appendix A: Glossary

Abiotic—not of or not relating to living organisms.

Activities (Forestry)—the action steps, tasks, procedures, and services performed in conjunction with implementing the management plan for you land.

Advanced Reproduction—young trees established before a regeneration cutting.

Aesthetics—that which appeals to the senses.

Age Class—a distinct aggregation of trees originating from a single natural event or regeneration activity. All trees in a stand within a given age interval, usually 10 or 20 years.

Amortization Deduction—associated with the Investment Tax Credit. The ability to capitalize 95 percent of the expenses associated with reforestation (up to $10,000) over an 84-month period.

Apical—at the apex. Refers to a leaf or bud at the tip of a stem, or the growth at the tip of a branch. This is where the plant spends most of its growth energy.

Artificial Regeneration—the establishment of a stand of trees through planting seedlings or cuttings, or by direct seeding.

Bareroot Seedling—a seedling lifted from a nursery with its roots freed from the soil in which it had been grown.

Basal Area—typically, the total cross section area (in square feet) of all the timber on an acre, at a height of 4.5 feet.

Best Management Practices (BMPs)—a practice or combination of practices determined to be the most effective, practical means of reducing pollution and other environmental impacts generated by forestry activities. In most states, BMPs have been codified into a set of regulations or laws that serve as the minimum requirements for silvicultural activities.

Biltmore Stick (or Cruiser's Stick)—a basic tool for measuring the board-foot volume of standing trees. It is a 25-inch long, calibrated measuring stick.

Biotic—of or relating to living organisms.

BMPs—(See **Best Management Practices**).

Board Foot—equivalent to a piece of lumber one inch thick by 12 inches wide by one foot long.

Browse—green vegetation or stems of woody plants used by wildlife for food.

Chopping—a method of site preparation where a bulldozer pulls a large (8-10' diameter) rolling drum with blades that cut up small trees and saplings in to 2-4' lengths. This gets these undesirable species down on the ground and, when dried out after several months, ready for a prescribed burn.

Clearcut—a method of regenerating an even-aged stand in which a new age class develops in a fully-exposed micro-environment after removal, in a single cutting, of all trees in the previous stand. Clearcutting is commonly used with shade intolerant species which require full sunlight to reproduce and grow well. Silvicultural clearcuts are those where all trees are harvested, as opposed to economic or commercial clearcuts where only the merchantable trees are removed.

Competition—in a forest, the struggle for light, nutrients and water among neighboring trees.

Container-Grown Seedlings—a seedling grown in a receptacle containing the soil or medium in which it has developed either from seed or as transplants.

Controlled Burning or Prescribed Burning—preplanned fire that is deliberately set in a time and manner to meet specific habitat or timber management objectives. Normally, it is designed to burn low growing brush and hardwoods, to consume excessive fuel on the ground, and to not harm the larger trees. Controlled burns should be planned and supervised by qualified professionals.

Coppice—(See **Sprout**).

Cord—a standard cord is a stack of wood that measures 128 cubic feet (eg. 4′ × 4′ × 8′). This includes bark, wood, and voids and is an indication of space occupied rather than actual wood measured.

Cost-Share Payment—a payment, usually in the form of a reimbursement to the landowner, for costs associated with forestry practices.

Critical Habitat—the geographical area that is deemed necessary for a particular species' survival at the time the species is listed under the Endangered Species Act. Typically, this is the area physically occupied by the species; however, in some cases the area can be enlarged to be the historic range of the species.

CRP (Conservation Reserve Program)—A federal government cost share program described in Chapter 8.

Cunit—equivalent to 100 cubic feet of solid wood (as opposed to stacked wood measured in cords).

DBH (Diameter at Breast Height)—the diameter of trees at a height of 4.5 feet above ground level.

Deed—a document that transfers rights associated with property from one person to another. A **Timber Deed** transfers rights to standing timber (usually when all timber within a prescribed boundary is to be cut) from the seller to the buyer.

Defect—in this context of Chapter 12, defect is anything that is not continuous sound wood (for example, mechanical damage during harvesting or processing, decay, and knots) that can impact log quality, and hence, calculated log weight.

Diameter-Limit Cutting—a harvest based on cutting all trees in the stand over a specified diameter, regardless of tree vigor, species or spatial distribution. Usually results in the long-term degradation of the stand.

Direct Seeding—the manual or mechanical sowing of tree seed on an area, either in spots or broadcast.

Doyle Log Rule—in the southern and eastern United States, the most commonly used log rule (the estimate of the number of board feet that can be sawn from a log). The formula for the Doyle log rule is $(D-4)2 \times L/16$, where D=diameter in inches of the log on the small end, inside bark and L = length of the log in feet. The Doyle Rule underestimates small logs and overestimates large logs.

Ecosystem—the system of interactions between living organisms and their environment.

Edge—an area where one type of habitat meets and blends with another. Usually associated with a more diverse plant community and habitat for many game species.

EFCRP (Emergency Forestry Conservation Reserve Program)—A federal government cost share program described Chapter 8.

Endangered Species—(1) animals and plants that will likely become extinct unless protected. Special management considerations are needed for these species; (2) an animal or plant listed under the Endangered Species Act.

Environmental Impact Statement (EIS)—a statement of the environmental effects of a proposed action and of alternative actions. Section 102 of the National Environmental Policy Act requires an EIS for all major federal actions.

Epicormic—shoot arising from an adventitious or dormant bud on a stem or branch of a woody plant, sometimes in response to infection or injury.

EQIP (Environmental Quality Incentives Program)—A federal government cost share program described in Chapter 8.

Escheat—reversion of property to the state in the absence of legal heirs or claimants; property that has reverted to the state when no legal heirs or claimants exist.

Extension Forester—a forester who works for the Cooperative Extension Service to develop and provide technical information for state forestry interests. The primary responsibility is group education, and includes preparing materials for local and regional educational activities.

Even-Aged Management—stand management designed to remove (harvest) all trees at one time, or over a short period, to produce even-aged (single age class) stands.

Face—an area on a graded tree one-quarter of the circumference and extending an entire 16 foot length. A **Clear Face** is a face free from knots that are more than one-half inch in diameter, or free from overgrown knots of any size, and free from holes over one-quarter inch in diameter.

Farm Service Agency (FSA)—The Farm Service Agency (FSA) administers and manages farm commodity, credit, conservation, disaster and loan programs as laid out by Congress through a network of federal, state and county offices. The Farm Service Agency is a part of the US Department of Agriculture. See www.fsa.usda.gov for more information.

Financial Analysis—the use of commonly accepted financial equations and concepts such as net present value, internal rate of return, benefit-cost and equal annual equivalent, to make decisions concerning the profitability of timber or other investments (hunting lease, recreation, etc).

FIP (Forestry Incentives Program)—A federal government cost share program described in Chapter 8.

Fire Break—a natural or artificial barrier usually created by removing vegetation to prevent or retard the spread of fire.

FLEP (Forest Land Enhancement Program)—A federal government cost share program described in Chapter 8.

Foliar—of or pertaining to a leaf, leaves, or foliage.

Forb—any herbaceous plant other than grass or legumes.

Forestland—land at least 10 percent stocked by forest trees of any size, or formerly having had such tree cover and not currently developed for non-forest use. The minimum area for classification of forestland is one acre and must be at least 100 feet wide. Forestland is distinguished from rangeland in transition vegetation types if the tree canopy cover exceeds 10 percent. Forestlands include cutover areas temporarily unstocked as well as young stands and plantations established for forestry purposes which do not yet have 10 percent crown cover. [source: USDA-SCS 1982 NRI]

Forest Products—generally, products whose main component originated from trees. Examples include lumber, paper, rayon, nail polish, linoleum, cellophane, paint and paint remover.

Forestry Consultant—a self-employed, sometimes registered forester who provides landowners with professional advice and services for a fee. Services range from handling timber sales to providing for or supervising every form of forest management imaginable. Some consultants specialize in wetlands, soils, wildlife biology, recreation, etc., while others offer general assistance.

Forest Type—groups of tree species commonly found growing in the same stand because their environmental requirements are similar. Examples of forest types are: oak-hickory, longleaf-slash pine, elm-ash-cottonwood and oak-gum-cypress, loblolly.

Friable—a soil consistency term pertaining to the ease of crumbling of soils.

Game Species—wildlife that is managed and hunted for recreation.

Gross Log Scale—the volume based on actual log dimensions. **Net Log Scale** is adjusted for defects based on the projected use of the logs. **Net Fiber Scale** is the Gross Scale adjusted for defects (voids, decay, charred wood, etc.) that reduce the amount of wood usable for pulping or other chip products. **Net Product Scale** has additional adjustments for defects (sweep, cracks, shake, etc.) that affect the yield of solid wood products such as lumber and veneer. Log volume can be reported in either gross or net scale basis; net scale is more common.

Habitat—a place where a plant or animal naturally lives that provides some or all of the needs for survival. Components of habitat include food, water, cover and living space.

High Grading—a harvesting technique that removes only the best trees to obtain high, short-term financial returns at the long-term expense of remaining stand growth potential.

Home Range—an area where an animal spends most of its time from birth until death and which provides most of the animal's habitat needs.

Hypo-Hatchet—a hatchet that injects a preset amount of herbicide or chemicals into tree stems on impact. (See **Tree Injector.**)

Intermediate Cuts—removing immature trees from the forest at a period between early growth and maturity to improve the quality of the remaining forest stand for timber, wildlife, or other objectives.

Intermediate Treatments—silvicultural operations performed in existing forest stands between regeneration and final harvest such as thinning, pruning, fertilization, and controlled burning, designed to improve the quality and growth of the remaining trees in the stand.

International Log Rule—a log rule used predominantly by the USDA Forest Service and to a lesser extent by the private sector. Overall, the International 1/4-Inch Log Rule is the most consistent.

Investment Tax Credit (ITC)—(See **Reforestation Tax Credit**).

Investor—for IRS purposes, a timberland investor whose involvement is minimal and whose purposes are to grow timber for capital appreciation.

Landowner Assistance Program—a program offered by a public or private entity which may involve various technical, financial or information forms of support.

Lease—an agreement between the landowner (leasor) and land user (leasee) which grants the land user rights for specified activities, such as hunting, on the property.

Log Rule—estimates the board foot volume that can be sawn from trees or logs. Commonly used log rules include Doyle, Scribner, and International.

Long-Term Capital Gain—the increase in value of a long term or large value asset—such as land and timber.

Lump Sum—a type of timber sale where buyers bid for the entire stumpage involved in a sale at a fixed price. (Contrast to **Per Unit.**)

Management Expense—expenses associated with managing your forested property.

Management Plan—usually a written document prepared by a natural resource professional that includes overall guidelines and recommended management practices to meet the owner's objectives for forest land. Includes maps of the property showing forest stands and other features, a time table for management practices, and a timber inventory.

Mast—the fruit of shrubs or trees that provides food for wildlife. Mast can be soft (such as blueberries) or hard (such as acorns).

Material Participant—for IRS purposes, a timberland investor whose involvement can be considered regular, continuous and substantial. The IRS has various tests for determining whether an investor is a Passive or Material Participant.

Moisture Content—the amount of water contained in wood. It is directly related to the temperature and humidity of surrounding air.

Multiple Use Management—land management for more than one purpose, typically for some combination of wood production, water, wildlife, recreation or aesthetics.

Natural Regeneration—the establishment of a stand of trees through seed or sprouts originating from trees already on the site.

Natural Resource Conservation Service (NRCS)—The Natural Resource Conversation Service (NRCS) conservation programs help people reduce soil erosion, enhance water supplies, improve water quality, increase wildlife habitat, and reduce damages caused by floods and other natural disasters. Public benefits include enhanced natural resources that help sustain agricultural productivity and environmental quality while supporting continued economic development, recreation, and scenic beauty. See *http://www.nrcs.usda.gov/* for more information.

Net Log Scale—(See **Gross Log Scale**).

Non-Game Wildlife—wildlife that is not hunted.

Nonpoint Source Pollution—water pollution from numerous widespread locations or sources that have no well-defined points of origin.

Objective—a management statement reflecting values or desired outcomes related to a vision of the future; (a function is simply a statement of an activity; an objective indicates the intention to achieve an outcome from the performance of the activity and should have both qualitative and quantitative attributes).

Passive Participant—for IRS purposes, a timberland owner whose involvement cannot be considered regular, continuous, and substantial. The IRS has various tests for determining whether an investor is a Passive or Material Participant.

Per Unit—a type of timber sale where the buyer pays the timber seller according to what is harvested on the site on a unit basis, e.g. per cord or per 1000 board feet. (Contrast to **Lump Sum.**)

Pesticide—a chemical substance (e.g. an insecticide, herbicide or fungicide) that kills harmful organisms and is used to control pests, such as insects, weeds or microorganisms.

Piling—in conjunction with shearing, the practice of pushing with a bulldozer, the downed saplings and small trees into piles for eventual burning, or into windrows where the vegetation will eventually rot.

Plantation—a reforested area established by planting trees.

Plastic Soil—a soil capable of being molded or deformed continuously and permanently, by relatively moderate pressure, into various shapes.

Practices (Forestry)—the techniques, methodologies, procedures, and processes that are used to achieve the objectives of a forestry management plan. (See **Activities**.)

Premerchantable Timber—a term used to define trees that have not yet reached a marketable value.

Prescribed Burning—a fire set by trained professionals under specific environmental conditions to obtain one or more forest management objectives. (See **Controlled Burning.**)

Present-Use Valuation—an assessment of the value of a parcel of land due to its current or present use rather than a potentially higher market value rate based on a potential future use.

Professional Forester—a person who has been professionally educated in forestry at a college or university.

Protective Regulatory (PR) Policies—policies and laws that safeguard society by limiting or mandating certain actions by the public and private sectors.

Reforestation—reestablishing a forest on an area where forest vegetation has been removed.

Reforestation Tax Credit (RTC)—provides, for each dollar of reforestation expense, a dollar for dollar reduction in the amount of taxes owed. The maximum federal tax credit for reforestation is $1,000.

Regeneration Cut—either the partial or complete harvest of trees to encourage the sprouting of desirable new tree growth.

Registered Forester—a person who annually registers and is licensed to practice forestry. Registration requirements include education, practical experience and examination. Renewal requirements include earning continuing education credits and abiding by established standards. Licensing or registering of foresters is not required by all states.

Renewable—resources able to be sustained or renewed indefinitely, because of new growth. (See **Sustainable**.)

Resource Assessment—a summarization of information gathered from an inventory of a particular piece of property. Data may include information on water, soil, timber, wildlife, aesthetics, recreation, etc.

Roadless Area—a national forest area which (1) is larger than 5000 acres, or if smaller than 5000 acres, contiguous to a designated wilderness or primitive areas; (2) contains no roads; and (3) has been inventoried by the Forest Service for possible inclusion in the wilderness preservation system.

Scribner Log Rule—a log rule used predominantly by the USDA Forest Service and to a lesser extent by the private sector. The Scribner Rule does not have an allowance for log taper and typically underestimates logs, particularly if the log length is long.

Seed Tree—(1) a tree left standing during stand harvest for the sole or primary purpose of providing seed to establish a new stand; (2) a method of natural regeneration which leaves just enough trees after a harvest to serve as a seed source for establishing the next stand under conditions of nearly full light exposure. Usually 5 to 10 trees per acre are retained. These are removed later after sufficient regeneration is established.

Selection—a method of creating new age classes in uneven-aged stands in which individual trees of all size classes or small groups of trees (group selection) are removed more or less uniformly throughout the stand to achieve desired stand structural characteristics. This method creates openings suitable for regeneration of shade tolerant species.

Shearing—the act of severing saplings and small trees near ground level by the forward motion of a blade on the front of a moving bulldozer.

Shelterwood—a method of regenerating an even-aged stand in which a new age class develops beneath the moderated micro-environment provided by residual trees. Trees are harvested in two or more operations. New seedlings grow and become established in partial shade protection of older trees. Harvests are usually 5 to 10 years apart, yielding an even-aged stand.

Shelterwood with Reserves—a variant of the shelterwood method in which some or all of the reserve (shelter) trees are retained after regeneration has become established to attain goals other than regeneration.

SIP (Stewardship Incentives Program)—A federal government cost share program described in Chapter 8.

Site—an area evaluated for its capacity to accomplish some objective. Evaluation is based on combined biological, climatic and soil factors.

Site Index—a species-specific measure of actual or potential forest productivity, expressed in terms of the average height of the dominant trees in a specified stand component (defined as the largest and tallest trees in a stand) at a specified index or base age. For example, site index 65 (age 50) means the dominant trees in a particular stand will be 65 feet tall at age 50. The higher the number the better the site.

Site Preparation—manual, mechanical, or chemical manipulation of a harvested site intended to enhance regeneration success.

Site Quality—the productive capacity of a site, usually expressed as volume production of a given species.

Snag—a standing dead or dying tree sometimes left remaining in a forest stand for wildlife perching, feeding, nesting sites or dens.

Soil Consistence—the combination of properties of soil material that determine its resistance to crushing and its ability to be molded or changed in shape. Such terms as loose, friable, firm, soft, plastic and sticky describe soil consistence.

Soil Drainage—the frequency and duration of periods when the soil is free of saturation with water.

Soil Fertility—the status of a soil with respect to the amount and availability to plants of elements necessary for plant growth.

Soil Survey—the systematic examination, description, classification, and mapping of soils in an area.

Soil Texture—the relative proportions of the various soil separates in a soil.

Soil Textural Class—a grouping of soil textural units based on the relative proportions of the various soil separates (sand, silt, and clay). These textural classes, listed from the coarsest to the finest in texture, are sand, loamy sand, sandy loam, loam, silt loam, silt, sandy clay loam, clay loam, silty clay loam, sandy clay, silty clay, and clay. There are several subclasses of the sand, loamy sand, and sandy loam classes based on the dominant particle size of the sand fraction (e.g., loamy fine sand, coarse sandy loam). (See chapter 5, table 5.2).

Specific Gravity—a comparison of density of a substance to that of water. Therefore, for wood density, both specific gravity and moisture content are important.

Sprout—a stem vegetatively produced from a stump, stem, or the roots. A new growth of a plant, such as a new branch or bud. (Similar to **Coppice.**)

Stand Description—a textual description that defines characteristics of a group of trees that are similar in composition. (See **Stand.**)

Stand (Forestry)—a group of trees similar enough to allow treatment as a single unit in a forest management plan.

Stand Map—a graphical description that shows the location of various stands on a specific piece of property. (See **Stand.**)

Stand Number—a number depicted in a table, map or text that refers to a specific stand on a specific piece of property. (See **Stand.**)

Stand Structure—the horizontal and vertical distribution of components of a forest stand including height, diameter, crown layers and stems of trees, shrubs, herbaceous understory, snags and down woody debris.

State Forestry Agency Forester—a forester employed by a state forestry agency to provide forestry services and to supervise state and federal forestry programs.

Streamside Management Zones (SMZs)—a strip of vegetation along ponds, streams and rivers which is left intact when adjacent forests are harvested. These areas are left intact because of their value in preventing soil erosion, maintaining water quality, and enhancing wildlife habitat.

Sustainable—the yield of a natural resource that can be produced continually from generation to generation, without depleting the resource. (See **Renewable.**)

Stumpage—(1) the value of timber as it stands uncut in terms of an amount per unit area; (2) a term for standing timber.

Subsoil—that part of the soil below the topsoil or "plow" layer. (See **Topsoil.**)

Sweep—the amount of curvature in a tree or log when compared to a flat plane.

Thinning—cutting trees in a forest stand to reduce stocking density (number of trees per acre) which will concentrate site productivity on fewer, higher quality trees.

Timber—a stand of trees suitable for sawing into lumber.

Timber Cruise—a survey of forestland to estimate species, stocking, volumes, products, size, and quality levels of standing timber. Cruise information is used for selling your timber, harvesting, and general management.

Timber Stand Improvement—treatments applied to an existing stand of timber for the purposes of improving stand composition, structure, condition, health, and/or growth.

Topsoil—the fertile uppermost layer of soil typically moved in cultivation. (See **Subsoil.**)

Tract—a parcel of land considered separately from adjoining land because of differences in ownership, timber type, management objectives, or other characteristics.

Transpiration—the emission of water vapor from the leaves of plants.

Tree Injector—an instrument used to inject a chemical into the wood of a tree. The injected substance may be a herbicide to kill unwanted trees, or a chemical to control disease. (See **Hypo-Hatchet.**)

Two-Aged Management—methods designed to maintain and regenerate a stand with two age classes. The resulting stand may be two-aged or tend towards an uneven-aged condition as a consequence of both an extended period of regeneration establishment and the retention of reserve trees that may represent one or more age classes.

Uneven-Aged Management—managing a forest by periodically harvesting trees of all ages to maintain a broad age (or size) class distribution. The forester maintains a greater number of trees in each smaller age-class than in the next older or larger class, up to some maximum age. This type of management is not common and is difficult to accomplish in southern hardwoods.

Watershed—the area of land from which rainfall (and/or snow melt) drains into a single point. Watersheds are also sometimes referred to as drainage basins or drainage areas. Ridges of higher ground generally form the boundaries between watersheds. At these boundaries, rain falling on one side flows toward the low point of one watershed, while rain falling on the other side of the boundary flows toward the low point of a different watershed. Large watersheds, like the Mississippi River basin contain thousands of smaller watersheds.

WHIP (Wildlife Habitat Incentives Program)—A federal government cost share program described in Chapter 8.

Wildfire—an unplanned or unwanted natural or human-caused fire.

Wildlife Management Plan—written plan detailing how and when to implement wildlife habitat and population improvement practices.

Windthrow—in forestry, windthrow refers to trees uprooted by wind, or to the phenomenon that causes such uprooting to occur. Windthrow commonly results from logging when the removal of certain trees at a forest's edge increases the exposure of the remaining trees to the wind. The resulting damage can be a significant factor in the evolution of a forest. Most commonly occurs among shallow-rooted species on sites with sallow soils, and in areas where cutting has reduced the density of a stand, exposing residual trees to the wind and depriving them of the accustomed support of adjacent trees.

Windrow (Forestry)—a line of organic material (remaining after a timber harvest) that is left on the ground to dry and decompose.

Wood Dealers—usually third party timber buyers who buy rights to standing timber from landowners and merchandise the stumpage products to the highest value uses.

WRP (Wetlands Reserve Program)—A federal government cost share program described in Chapter 8.

Appendix B:

State Extension Forest Resources

Belle W. Baruch Institute of Coastal Ecology & Forest Science
PO Box 596
Georgetown, SC 29442
TEL: 843-222-6323
FAX: 843-546-6296

Clemson University
Department of Forestry & Natural Resources
262 Lehotsky Hall
Clemson, SC 29634-1003
TEL: 864-656-3302
FAX: 864-656-0321

University of Tennessee
Dept. of Forestry, Wildlife & Fisheries
Knoxville, TN 37996-4563
TEL: 865-974-7346
FAX: 865-974-4714

Tennessee State University
Cooperative Agriculture Research Programs
3500 John Merrit Blvd.
Farrell/Westbrook Complex
Nashville, TN 37209-1561
TEL: 615-963-5616
FAX: 615-963-5833

Louisiana State University
Department of Agricultural Economics & Agribusiness
242A Agricultural Administration Building
Baton Rouge, LA 70803
TEL: 225-578-0344
FAX: 225-578-2716

Texas A&M University
Texas Cooperative Extension
301 Horticulture/Forest Science Building
College Station, TX 77843-2135
TEL: 979-845-1351
FAX: 979-458-6049

North Carolina State University
College of Natural Resources
Campus Box 8003
Raleigh, NC 27695-8003
TEL: 919-515-5574
FAX: 919-515-6883

Virginia Tech
College of Natural Resources
324 Cheatham Hall
Blacksburg, VA 24061
TEL: 540-231-9759
FAX: 540-231-8868

Virginia Tech
Albermarle County Extension Office
460 Stagecoach Road
Charlottesville, VA 22902
TEL: 434-872-4580
FAX: 434-872-4578

University of Arkansas–Monticello
School of Forest Resources
PO Box 3468
Monticello, AR 71656
TEL: 870-460-1052
FAX: 870-460-1092

University of Arkansas
Cooperative Extension Service
PO Box 391
Little Rock, AR 72203
TEL: 501-671-2346
FAX: 501-671-2185

Mississippi State University
Department of Forestry
Box 9681
Mississippi State, MS 39762
TEL: 662-325-8003
FAX: 662-325-8726

University of Florida
School of Forest Resources and Conservation
PO Box 110410
Gainesville, FL 32611-0410
TEL: 352-846-0891
FAX: 352-846-1277

Oklahoma State University
Department of Natural Resource Ecology & Management
008C Agriculture Hall
Stillwater, OK 74078
TEL: 405-744-8065
FAX: 405-744-3530

Auburn University
School of Forestry
M. White Smith Hall
Auburn University, AL 36849-5628
TEL: 334-844-1044
FAX: 334-844-1084

University of Kentucky
Department of Forestry
213 Thomas Poe Cooper Building
Lexington, KY 40546-0073
TEL: 859-257-5994
FAX: 859-323-1031

The University of Georgia
Warnell School of Forestry & Natural Resources
4-404 Forest Resources Building
Athens, GA 30602
TEL: 706-542-2866
FAX: 706-542-3342

This contact list was provided by the Southern Regional Extension Forestry web site. For more information on these State Extension Forest contacts and lists of other resources, visit the Southern Regional Extension Forestry web site: *www.sref.info.* Please note that the contact information is subject to change. Please reference the web site noted above to verify current addresses and phone numbers.

For more contacts, including additional states, visit *http://www.csrees.usda.gov/nea/nre/pdfs/forest_directory.pdf.* (If this link does not work, go to *http://www.csrees.usda.gov/* and in the "More Quick Links" box on the right click on "Programs", then select "Forests". Toward the bottom of the Forests page, under the "Resources" header, click on the link "Forestry and Forest Products Extension Directory. A national list of forestry and forest products extension professionals." This will bring up the list: "Cooperative Extension System Personnel in Forestry & Forest Products".)

Appendix C:
State Forestry Agencies

USDA Forest Service—Southern Research Station—SGSF
Jackson Drive
Gray, GA 31032
TEL: 478-986-5065

Alabama Forestry Commission
513 Madison Avenue
Montgomery, AL 36130
TEL: 334-240-9304
FAX: 334-240-9388

Oklahoma Department of Agriculture—Forestry Services
2800 North Lincoln Blvd.
Oklahoma City, OK 73105
TEL: 405-522-6158
FAX: 405-522-4583

Louisiana Office of Forestry
PO Box 1628
Baton Rouge, LA 70821
TEL: 225-952-8002
FAX: 225-922-1356

Virginia Department of Forestry
900 Natural Resources Drive
Suite 800
Charlottesville, VA 22903
TEL: 434-977-6555
FAX: 434-296-2369

Texas Forest Service
College Station, TX 77843-2136
TEL: 409-845-2601
FAX: 409-862-2463

Florida Division of Forestry
3125 Conner Blvd.
Tallahassee, FL 32399-1650
TEL: 850-488-4274
FAX: 850-488-0863

Kentucky Division of Forestry
627 Comanche Trail
Frankfort, KY 40601
TEL: 502-564-4496
FAX: 502-564-6553

Mississippi Forestry Commission
301 N. Lamar Street, Suite 300
Jackson, MS 39201
TEL: 601-359-2801
FAX: 601-359-1349
URL: http://www.mfc.state.ms.us/

South Carolina Forestry Commission
PO Box 21707
Columbia, SC 29221
TEL: 803-896-8800
FAX: 803-896-8097

Tennessee Department of Agriculture
Forestry Division
PO Box 40627
Melrose Station, TN 37204
TEL: 615-837-5411
FAX: 615-837-5003

Arkansas Forestry Commission
3821 West Roosevelt Rd
Little Rock, AR 72204-6396
TEL: 501-296-1941
FAX: 501-296-1949
URL: http://www.forestry.state.ar.us/

North Carolina Division of Forest Resources
1616 Mail Service Center
Raleigh, NC 27699
TEL: 919-733-2162
FAX: 919-715-4350
URL: http://www.dfr.state.nc.us/

Georgia Forestry Commission
PO Box 819
Macon, GA 31202-0819
TEL: 478-751-3500
800-GATREES
FAX: 478-751-3465
URL: http://www.gfc.state.ga.us/

Southern Group of State Foresters
PO Box 930
Winder, GA 30680
TEL: 770-267-9630
FAX: 425-920-1661
URL: http://www.southernforests.org/

This contact list was provided by the Southern Regional Extension Forestry web site. For more information on these State Forestry Agencies and lists of other resources, visit the Southern Regional Extension Forestry web site: *www.sref.info*. Please note that the contact information is subject to change.

Appendix D:
State Wildlife Contacts

Alabama Wildlife and Fresh Water Fisheries Division
64 N. Union Street, Suite 468
Montgomery, Alabama 36130
TEL: 334-242-3465
FAX: 334-242-3489
URL: www.outdooralabama.com/

Arkansas Game and Fish Commission
2 Natural Resources Drive
Little Rock, AR 72205
TEL: 501-223-6300
800-364-4263
FAX: 501-676-2734
URLS: www.agfc.com/contact/
www.agfc.com/

Florida Fish and Wildlife Commission
620 S. Meridian St.
Tallahassee, FL 32399-1600
TEL: 850-488-4676
FAX: 850-413-0381
URL: http://myfwc.com/

Georgia Wildlife Resources
2070 U.S. Hwy. 278, S.E.
Social Circle, GA 30025
TEL: 770-918-6400
FAX: 706-557-3030
URLS: https://georgiawildlife.dnr.state.ga.us/service/supportemail.asp
http://georgiawildlife.dnr.state.ga.us/

Kentucky Department of Fish and Wildlife Resources
1 Sportsman's Lane
Frankfort, KY 40601
TEL: 800-858-1549
FAX: 502-564-0506
URL: http://fw.ky.gov/

Lousiana Department of Wildlife and Fisheries
2000 Quail Drive
Baton Rouge, LA 70808
TEL: 225-765-2800
FAX: 225-765-0948
URLS: www.wlf.state.la.us/contactus/
www.wlf.state.la.us/

Mississippi Department of Wildlife, Fisheries and Parks
1505 Eastover Drive
Jackson, MS 39211-6374
TEL: 601-432-2400
FAX: (Please contact agency for departmental fax numbers.)
URLS: www.mdwfp.com/CustomerService/Enter.asp
www.mdwfp.com/default.asp

North Carolina Division of Marine Fisheries
3441 Arendell Street
Morehead City, NC 28557
TEL: 252-726-7021
800-682-2632
FAX: 252-727-4828
URLS: www.ncfisheries.net/content/cont1/contactdmf.htm
www.ncfisheries.net/index.html

North Carolina Wildlife Resources Commission
1751 Varsity Drive,
Raleigh, NC 27606
TEL: 919-707-0010
FAX: 919-707-0067
URL: www.ncwildlife.org/

Oklahoma Department of Wildlife Conservation
1801 N Lincoln
Oklahoma City, OK 73105
TEL: 405-521-3851
FAX: 405-521-6535
URL: www.wildlifedepartment.com//

South Carolina Department of Natural Resources
1000 Assembly Street
Columbia, SC 29201
TEL: 803-734-9100
FAX: 803-734-9200
URLS: www.dnr.sc.gov/admin/divphone.html
www.dnr.sc.gov/

Tennessee Wildlife Resources Agency
PO Box 40747
Nashville, TN 37204
TEL: 615-781-6500
FAX: (Please contact agency for departmental fax numbers.)
URL: www.state.tn.us/twra/

Texas Parks and Wildlife
4200 Smith School Road
Austin, TX 78744
TEL: 800-792-1112
FAX: (Please contact agency for departmental fax numbers.)
URL: www2.tpwd.state.tx.us/business/feedback/webcomment/?p=%252Findex.phtml
http://www.tpwd.state.tx.us/

Texas Wildlife Damage Management Service
P.O. Box 100410
San Antonio, Texas 78201-1710
TEL: 210-472-5451
FAX: 210-472-5446
URL: http://agextension.tamu.edu/twdms/twdmshom.htm

Virginia Department of Game and Inland Fisheries
4010 W. Broad St.
Richmond, Virginia 23230
TEL: 804-367-1000
FAX: 804-367-9147
URL: www.dgif.state.va.us/

Virginia Marine Resources Commission
2600 Washington Avenue, 3rd Floor
Newport News, VA 23607
TEL: 757-247-2200
FAX: 757-247-2020
URL: www.mrc.state.va.us/

This contact list was provided by the Southern Regional Extension Forestry web site. For more information on these State Wildlife Contacts and lists of other resources, visit the Southern Regional Extension Forestry web site: www.sref.info. Please note that the contact information is subject to change.

Appendix E:

State Forestry Associations

Mississippi Forestry Association
201 Realtors Building
620 North State Street
Jackson, MS 39202
TEL: 662-354-4936
FAX: 662-354-4937

Kentucky Forest Industries Association
106 Progress Drive
Frankfort, KY 40601
TEL: 502-695-3979
FAX: 502-695-8343

Tennessee Forestry Association
Box 290693
Nashville, TN 37229
TEL: 615-883-3832
FAX: 615-883-0515

Oklahoma Forestry Association
Box 238
Idabel, OK 74745
TEL: 580-286-3970
FAX: 580-286-9351

Virginia Forest Products Association
PO Box 72080
Richmond, VA 23255-2080
TEL: 804-741-0836

Texas Forestry Association
Box 1488
Lufkin, TX 75902-1488
TEL: 409-632-8733
FAX: 409-632-9461

Alabama Forestry Association
555 Alabama Street
Montgomery, AL 36104-4395
TEL: 334-265-8733
FAX: 334-262-1258

Forest Landowners Association
900 Circle 75 Parkway, Suite 205
Atlanta, GA 30339
TEL: 404-325-2954
800-325-2954
FAX: 404-325-2955
URL: http://www.forestlandowners.com/

Alabama Forest Owners' Association, Inc
Box 361434
Birmingham, AL 35236
TEL: 205-987-8811
FAX: 205-987-9824

Georgia Forestry Association, Inc.
551 North Frontage Road
Forsyth, GA 31029
TEL: 478-992-8110
800-9-Grow-GA
FAX: 478-992-8109
URL: http://www.gfagrow.org/

Arkansas Forestry Association
410 South Cross Street #C
Little Rock, AR 72201
TEL: 501-374-2441
FAX: 501-374-6413

South Carolina Forestry Association
Box 21303
4901 Broad River Road
Columbia, SC 29221
TEL: 803-798-4170
FAX: 803-798-2340

North Carolina Forestry Association
1600 Glenwood Avenue, Ste. 1
Raleigh, NC 27608-2355
TEL: 919-834-3943
FAX: 919-832-6188

Louisiana Forestry Association
PO Box 5067
Alexandria, LA 71307
TEL: 318-443-2558

Virginia Forestry Association
3808 Augusta Avenue
Richmond, VA 23230-3910
TEL: 804-278-8733
URL: www.vaforestry.org

Florida Forestry Association
PO Box 1696
Tallahassee, FL 32302
TEL: 850-222-5646
FAX: 850-222-6179
URL: www.floridaforest.org

This contact list was provided by the Southern Regional Extension Forestry web site. For more information on these State Forestry Association contacts and lists of other resources, visit the Southern Regional Extension Forestry web site: *www.sref.info*. Please note that the contact information is subject to change.

Appendix F:

Southern Forestry Schools

Louisiana Tech University
PO Box 10138
Ruston, LA 71272-0045
TEL: 318-257-3392
FAX: 318-257-5061

Texas A & M University
Department of Ecosystem Science and Management
College Station, TX 77843-2135
TEL: 979-845-5000

Stephen F. Austin State University
School of Forestry
Box 6109, SFA Station
Nacogdoches, TX 76962-6109
TEL: 409-568-3304
FAX: 409-568-2489

The University of Florida
349 Newins-Ziegler Hall
PO Box 110410
Gainesville, FL 32611-0410
TEL: 352-846-0845
FAX: 352-392-1277

Auburn University
School of Forestry and Wildlife Sciences
108 M. White Smith Hall
Auburn, AL 36849-5418
TEL: 334-844-1007
FAX: 334-844-1084

The University of Kentucky
Department of Forestry
Lexington, KY 40546-0073
TEL: 859-257-7596
FAX: 859-323-1031
URL: http://www.uky.edu/Agriculture/Forestry/forestry.html

Virginia Tech University
Department of Forestry
307 Cheatham Hall (0324)
Blacksburg, VA 24061
TEL: 540-231-6952
FAX: 540-231-3698

University of Puerto Rico–Mayaguez Campus
Departamento de Agronima y Suelos
PO Box 5800
College Station
Mayaguez, PR 00681
TEL: 787-265-3851
FAX: 787-265-0860

North Carolina State University
Department of Forestry
Box 8008
Raleigh, NC 27695-8008
TEL: 919-515-4471
FAX: 919-515-6193

Oklahoma State University
018 Agricultural Hall
Stillwater, OK 74078
TEL: 405-744-5443
FAX: 405-744-3530

Mississippi State University
College of Forest Resources
Box 9680
Mississippi State, MS 39762
TEL: 662-325-2953
FAX: 662-325-8726

North Carolina State University
College of Natural Resources
2028 Biltmore Hall
Campus Box 8001
Raleigh, NC 27695-8001
TEL: 919-515-2883
FAX: 919-515-7231

Virginia Tech University
College of Natural Resources
324 Cheatham Hall
Blacksburg, VA 24061-0324
TEL: 540-231-3479
FAX: 540-231-7664

Louisiana State University
School of Renewable Natural Resources
227 Renewable Natural Resources Building
Baton Rouge, LA 70803-6200
TEL: 225-578-4131
FAX: 225-578-4227

University of Arkansas-Monticello
School of Forest Resources
PO Box 3468
Monticello, AR 71656
TEL: 870-460-1052
FAX: 870-460-1092

North Carolina State University
Department of Wood and Paper Science
Box 8005
Raleigh, NC 27695-8005
TEL: 919-515-5726
FAX: 919-515-6302

Clemson University
Department of Forestry and Natural Resources
262 Lehotsky Hall
Clemson, SC 29634-1003
TEL: 864-656-3302
FAX: 864-656-0321

Mississippi State University
Department of Forest Products
Box 9820
Mississippi State, MS 39762-9820
TEL: 662-325-4444
FAX: 662-325-8126

Mississippi State University
Department of Wildlife & Fisheries
Box 9690
Mississippi State, MS 39762
TEL: 662-325-2615
FAX: 662-325-8726

Auburn University
210 M White Smith Hall
Auburn University, AL 36849-5418
TEL: 334-844-1054
FAX: 334-844-1084

Mississippi State University
Department of Forestry
Box 9681
Mississippi State, MS 39762
TEL: 662-325-2681
FAX: 662-325-8726

Virginia Tech
Department of Fisheries & Wildlife Science
100 Cheatham Hall
Mail Code 0321
Blacksburg, VA 24061-0321
TEL: 540-231-5919
FAX: 540-231-7580

Duke University
School of Environment and Earth Sciences
A246 Levine Science Research Ctr.
Box 90328
Durham, NC 27708-0329
TEL: 919-613-8004
FAX: 919-684-8741

Mississippi State University
Department of Forestry
Box 9680
Mississippi State, MS 39762
TEL: 662-325-2696
FAX: 662-325-8726

The University of Georgia
4-406 Forest Resources Building
The University of Georgia
Athens, GA 30602
TEL: 706-542-2866
FAX: 706-542-3342

The University of Georgia
Daniel B. Warnell School of Forestry and Natural Resources
Athens, GA 30602-2152
TEL: 706-542-4741
FAX: 706-542-2281

North Carolina State University
Dept. of Parks, Rec. & Tourism Mgmt.
Box 8004
Raleigh, NC 27695
TEL: 919-515-3276
FAX: 919-515-3687

Texas A&M University
Department of Rangeland Ecology and Management
2126 TAMU
Texas A&M University
College Station, TX 77843-2126
TEL: 969-845-5579
969-845-6430

The University of Florida
School of Forest Resources and Conservation
138 Newins Ziegler Hall
Gainesville, FL 32611-0410
TEL: 352-846-0850
352-846-1277

University of Tennessee
Department of Forestry, Wildlife and Fisheries
274 Ellington Plant Sci. Bldg.
Knoxville, TN 37996-4563
TEL: 865-974-7126
865-974-7988
FAX: 865-974-4714

Virginia Tech
Dept. of Wood Science & Forest Products
210 Cheatham Hall
Blacksburg, VA 24061-0323
TEL: 540-231-8853
FAX: 540-231-8176

This contact list was provided by the Southern Regional Extension Forestry web site. For more information on these Southern Forestry Schools and lists of other resources, visit the Southern Regional Extension Forestry web site: *www.sref.info* and *www.napfsc.org*. Please note that the contact information is subject to change.

Appendix G:

State Natural Resource Agencies

Alabama Department of Conservation and Natural Resources
64 N. Union Street, Suite 468
Montgomery, Alabama 36130
TEL: 334-242-3486
FAX: 334-242-3489
URL: www.outdooralabama.com/

Arkansas Department of Environmental Quality
8001 National Drive
Little Rock, AR 72209
P.O. Box 8913
Little Rock, AR 72219
TEL: 501-682-0744
FAX: 501-682-0880 Public Outreach Division (Please include information on fax regarding area of inquiry)
URL: www.adeq.state.ar.us/

Florida Department of Environmental Protection
3900 Commonwealth Blvd. M.S. 49
Tallahassee, FL 32399
TEL: 850-245-2118
FAX: 850-245-2128
URL: www.dep.state.fl.us/

Georgia Department of Natural Resources
2 Martin Luther King, Jr. Drive, S. E., Suite 1252
East Tower
Atlanta, GA 30334
TEL: 404-656-3500
FAX: 404-656-0770
URLS: www.gadnr.org/email/index.php
www.gadnr.org/

Kentucky Department for Environmental Protection
14 Reilly Road
Frankfort, KY 40601
TEL: 502-564-2225
FAX: 502-564-4245 Fax
URL: www.dep.ky.gov/

Lousiana Department of Environmental Quality
602 N. Fifth Street
Baton Rouge, LA 70802
TEL: 225-219-5337
URLS: www.deq.louisiana.gov/portal/tabid/62/Default.aspx
(Please refer to this website to locate fax number for individual departments.)
www.deq.louisiana.gov/portal/

Lousiana Department of Natural Resources
P.O. Box 94396
Baton Rouge, LA 70804-9396
TEL: 225-342-4500
URLS: http://dnr.louisiana.gov/staff.ssi
(Please refer to this website to locate fax number for individual departments.)
http://dnr.louisiana.gov/

Mississippi Department of Environmental Quality
P. O. Box 20305 Jackson
MS 39289-1305
TEL: 601-961-5171
TEL: 888-786-0661
FAX: 601-354-6938
URLS: www.deq.state.ms.us/newweb/MDEQDirNew.nsf?OpenDatabase
www.deq.state.ms.us/MDEQ.nsf/page/Main_Home?OpenDocument

North Carolina Department of Environmental and Natural Resources
1601 Mail Service Center
Raleigh, NC 27699-1601
TEL: 919-733-4984
877-623-6748 Customer Service
FAX: 919-715-3060
919-715-7468 Customer Service
URL: www.enr.state.nc.us/

Oklahoma Department of Environmental Quality
707 N Robinson
Oklahoma City, OK 73102
P.O. Box 1677
Oklahoma City, OK 73101-1677
TEL: 405-702-1000
FAX: 405-702-1001
URL: www.deq.state.ok.us/

South Carolina Department of Health and Environmental Control
2600 Bull Street
Columbia, SC 29201
TEL: 803-898-3432
FAX: (Please contact agency for departmental fax numbers.)
URLS: www.deq.state.va.us/webmaster.html
www.scdhec.net/administration/contactus.asp
www.scdhec.net/

South Carolina Department of Natural Resources
1000 Assembly Street
Columbia, SC 29201
TEL: 803-734-9100
FAX: 803-734-9200
URLS: www.dnr.sc.gov/admin/divphone.html
www.dnr.sc.gov/

Tennessee Department of Environment and Conservation
401 Church Street L & C Annex, 1st Floor
Nashville, Tennessee 37243-0435
TEL: 888-891-8332
FAX: 615-687-7072
URL: www.state.tn.us/environment/

Texas Natural Resource Conservation Commission
12100 Park 35 Circle
Austin, TX 78753
P.O. Box 13087
Austin, Texas 78711-3087
TEL: 512-239-1000
512-239-4000 Public Assistance
FAX: 512-239-4007 Public Assistance
URL: www.tceq.state.tx.us/

Virginia Department of Environmental Quality
629 East Main Street
Richmond, VA 23218
P.O. Box 1105
Richmond, VA 23218
TEL: 804-698-4000
800-592-5482
FAX: 804-698-4500
URLS: www.deq.state.va.us/webmaster.html
www.deq.state.va.us/

This contact list was provided by the Southern Regional Extension Forestry web site. For more information on these State Natural Resource Agency Contacts and lists of other resources, visit the Southern Regional Extension Forestry web site *www.sref.info*. Please note that the contact information is subject to change.

Appendix H:

How to Choose a Consulting Forester

It is profitable to grow timber, but constant changes make the job more and more complex. Increasingly, there are tax, legal and environmental requirements. Products and measurements change and prices fluctuate. New management practices and techniques evolve. Ownership profiles are changing, and timber supply is not the issue it once was. For the past 50 years we have grown more timber than we have cut and the projections are for the next 50 years that we will continue this trend. All of this tells us that forestry is a specialized and competitive business, and management will be the key to profit in the future. It is shortsighted to seek help only when making a timber sale, or worse, after making a sale! The harvest is only part of the management plan, and several years worth of prescribed burning before a timber sale may allow for natural reseeding and eliminate costly site preparation and planting.

Numerous studies have shown increased returns will more than offset the consultant's fee. Methods of setting fees differ among consultants. Some perform all work on a time-charge basis (hourly) or daily rate plus expenses. Others quote fixed, per acre rates for cruising, prescribed burning, planting, timber marketing and other services, or per mile rates for surveying or boundary line maintenance. Many timber sales are handled on a percentage of income. Some consultants charge for management work only when a sale is made, while others prefer to bill regularly for management work. The fee method can usually be tailored to suit the client.

The less you know about timber and the less time you can spend on your timberlands, the more you need a consultant.

What Is A Consultant?

Think of a consultant as a full-time, independent forester whose services are available to the public and who does not work for a timber buying business. The highest standards of education, experience and ethics are set by the Association of Consulting Foresters (ACF), and ACF membership is one measure of professional standing. There are also well-qualified consultants who are not ACF members. In addition, there are many professional foresters who consult part-time, or during retirement or as a sideline to regular employment.

Forest industry and timber dealers often have foresters who assist owners (assistance foresters) and they sometimes use the term consultant. Industry assistance foresters may provide services for little or no direct charge, but hope to purchase an owner's timber. In working with anyone other than a full-time independent consultant, you should be alert to other interests, which may shape opinions or color his or her judgement. It is difficult for timber buyers, for instance, to serve landowners as objective counselors concerning timber cutting and marketing. A full-time independent consultant, however, works solely for the client's interest for a fee agreed upon in advance.

What Can He or She Do?

You can find a consultant to perform almost any forestry service. He or she can represent you on a sensitive issue such as a boundary line dispute, appraisal or litigation, can arrange site preparation, planting or roadwork, or can take over forest management completely and act for you in all matters. Consultants are especially well versed in timber sales, which for you, the owner, is the payoff point in management. Consulting foresters know about marketing, merchandising and seeking competitive bids. They know which contract terms are appropriate such as penalty rates, performance deposits and damage clauses and can give you full representation in any timber sale dispute. Some can also advise on tax treatment. There are both specialists and generalists in the ranks of consulting forestry, and some function as a general contractor, pulling in each specialty only when needed. A consulting forester can help establish property lines, assure that tax assessment and deed records agree, set up basic records, set goals, help you understand your land and timber conditions, list the things that should be done, evaluate alternatives, select, schedule and carry out work, and keep records of income, expense and volume.

In the area of financial management, consultants can provide appraisals, investment counsel, advice and analysis, record keeping for tax or other purposes, estate planning and assistance with timberland loans. In connection with litigation problems, they are frequently called on to assist in preparation and expert testimony for contract, tax, possession or trespass cases.

They are also available to handle one-time specialty problems, such as growing Christmas trees or making an equitable property division. Because they are helpful in so many areas, forestry consultants often become permanent advisors or managers for landowners.

How To Choose A Consultant

Consultants are individuals, and as in every profession, there are good ones and bad ones, ones that suit you and ones that do not. You must find one you like who is properly equipped, experienced and located for your needs. Making a list of available consultants is the first step. Some sources for this information are:

- Professional associations such as the Association of Consulting Foresters, (www.acf-foresters.com) or Society of American Foresters.
- Friends or neighbors who have employed a consultant.
- State forestry associations or forest industry organizations
- State or federal government sources such as a state forester, State Board of Registration for foresters, extension forester or Natural Resources Conservation Service forester.
- Attorneys, accountants or bankers who may have worked with consultants.

A consultant's expertise is what he has to sell, and education, training and experience have shaped his knowledge and judgement. He should have been in business long enough to have a proven track record. Does he have good professional background and keep up-to-date through continuing education courses, conferences and professional journals?

Also important is the staff. Some consulting firms employ foresters, technicians or laborers to carry out the field work. The quality of these employees is important. Are they permanent or temporary employees, and is the work well supervised?

In addition, your consultant needs to be located near a small job or perhaps handling several in the same area. Large jobs may be economical at any distance. Sometimes it is convenient for the consultant to be located near the owner instead of near the property. Is location important to you?

What about general philosophy? Consultants do not all think alike, and business and management philosophies may vary widely. There may be more than one good forest management practice for a given situation and a recommendation will depend on condition of the property, objectives of the owner and the forester's philosophy. Have you asked about forest management ideas?

Moreover, you should have a clear understanding concerning fees and work to be done. Will your consultant work on a time-charge basis, a fixed fee or percentage arrangement? Until a consultant is familiar with the tract, it may not be possible to predict costs accurately. Are you comfortable with the method for billing and the work agreement?

Above all, your consultant should be someone you can trust. Frequently a continuing personal relationship evolves. Do you feel he understands your objectives and that you can identify with him?

A Satisfactory Relationship

Once you select a consultant, there are certain things you can do to help make a long-lasting, satisfactory relationship. To make good recommendations your consulting forester must know your needs and wishes. Level with him or her telling clearly what you want. Often the hardest part of a consultant's job is getting the landowner to explain the objectives for the ownership. Then listen and let the consultant gather information which may be necessary. A doctor must examine the patient before diagnosing and prescribing, and must charge for the examination. When you have carefully chosen your consultant, trust becomes the key to your relationship. Make a commitment to your consultant, and trust the judgments given. He or she is dedicating a career to forestry advice and assistance. The ultimate goal is to serve you and your consulting forester's advice can be as valuable as that of your lawyer, banker or accountant.

Acknowledgments

This publication began as a correspondence course produced at North Carolina State University, Raleigh, NC, in 1986. The authors of this work were members of the Department of Forestry, North Carolina State University and the School of Forest Resources, North Carolina Agricultural Extension Service. The original publication can be found on the internet at:

http://sref.info/publications/online_pubs/woodland_correspondence.pdf/file_view

About 10 to 12 years later, the Forest Landowner Foundation requested donations to update this earlier correspondence course. The Foundation gratefully acknowledges the generous support of the organizations and individuals, listed in the Sponsors Acknowledgments section, who made the contributions which supported this work. We also appreciate the work of the Foundation members who assisted in raising the funds. During this period, the Foundation was headed by Mr. Carl Sewell and Dr. Shelton Short.

The update of the correspondence course was headed by Dr. Richard Brinker, Auburn University, and Mr. Rick Hamiltion, North Carolina State University, with coordination and other assistance provided by Mr. Bill Hubbard, Southern Regional Forester. We are grateful to them and those who assisted them.

After the update, the staff at the Forest Landowner Association assembled the material and provided additional editing. Notable among these contributors were Steve Newton and Page Cash. We thank them for moving the work along.

Later, the Forest Landowner Foundation continued the effort to publish the correspondence course. With the help of Bill Hubbard, SE Regional Forester, a team was assembled to revise and update the text (see the Chapter Authors and Editors section for the list of those indivduals). We appreciate their expertise, time and energy in completing this work. Many of them extensively revised their sections and added many pages of information. Their work greatly enhanced the completeness of this work.

Editing to bring the many chapters and sections into congruence was done by Charlie Finley of Verbatim Editing in Richmond Virginia. Charlie will always be appreciated for his contribution.

About this time, the decision was made to change the format from a correspondence course to a reference book with questions at each chapter's end. To make this change, Lisa Lester, my associate in Bellevue, Washington, spent many, many hours and contributed many of her talents to make this a consistent, uniform publication that includes all the necessary attributes of a first quality book. Lisa also served as the most reliable proof reader. We thank her for her many contributions in making this an easily readable book.

Sarah Sturm at the FLA office turned the manuscripts into print ready proofs, providing the artistry that gives the book its sense of style. Sarah contributed many long hours and much work in moving the text into the print ready proofs, as well as many hours working with Lisa in Bellevue to resolve an infinity of details. Our thanks to Sarah for all of her work, her artistic skill, and her time.

Reviewers Anna N. Smith, James G. Thwaite, and Dr. John Bowen and plus other trustees of the Forest Landowner Foundation have our thanks for their efforts in reviewing the text and providing their comments to improve the book.

Lastly, I thank my wife, Susan, for acting as a willing first editor and proof reader whenever I requested and for enduring my absence from family activities.

Philip A. Hardin, Ph.D.
Editor

Past President, Forest Landowner Foundation
Bellevue, Washington